Darwinism and Its Discontents

This book presents an ardent defense of Charles Darwin's theory of evolution against its many critics by one of the leading experts on this subject. Offering a clear and comprehensive exposition of the thinking of Darwin, Michael Ruse brings the story up to date, examining important issues such as the origins of life, the fossil record, the mechanism of natural selection, and rival theories such as punctuated equilibrium, the story of human evolution (including the recently found "hobbits," *Homo floresiensis*), fraud in biological science, literary approaches to evolution, and the philosophical and religious implications of Darwinism, notably a discussion of Creationism and its modern-day offshoot, Intelligent Design theory. Ruse draws on the most recent discoveries but writes with a minimum of jargon. His book will appeal to many readers, from professional biologists to concerned citizens who worry that Darwinism is a naturalistic religion that is forced on schoolchildren in the face of their own deeply held Christian convictions. Openly revealing his own beliefs, Ruse aims throughout to present information and critical tools so that readers can make informed decisions for themselves.

Michael Ruse is one of the world's leading authorities on the history and philosophy of Darwinian evolutionary theory. He is the author of many books; his most recent book with Cambridge University Press is *Can a Darwinian Be a Christian? The Relationship between Science and Religion* (2001). A Fellow of the Royal Society of Canada and of the American Association for the Advancement of Science, he has been a Herbert Spencer Lecturer at Oxford University and a Gifford Lecturer at Glasgow University, and he has also held Guggenheim and Isaak Walton Killiam fellowships.

Darwinism and Its Discontents

MICHAEL RUSE
Florida State University

CAMBRIDGE UNIVERSITY PRESS
Cambridge, New York, Melbourne, Madrid, Cape Town, Singapore, São Paulo

Cambridge University Press
32 Avenue of the Americas, New York, NY 10013-2473, USA

www.cambridge.org
Information on this title: www.cambridge.org/9780521829472

First published 2006

Printed in the United States of America

A catalog record for this publication is available from the British Library.

Library of Congress Cataloging in Publication Data

Ruse, Michael.
Darwinism and its discontents / Michael Ruse.
p. cm.
Includes bibliographical references (p.) and index.
ISBN 0-521-82947-X (hardback)
1. Evolution (Biology) 2. Darwin, Charles, 1809–1882. I. Title.
QH371.R755 2006
576.8′2 – dc22 2005031237

ISBN-13 978-0-521-82947-2 hardback
ISBN-10 0-521-82947-X hardback

For David Castle
Stephen Haller
Jean Lachapelle
Eduardo Wilner

Contents

Acknowledgments

I have been thinking about the ideas in this book for about four decades and have decided that the time has come to put them all together. In a way, this book is a kind of prequel (as they call it in the cinema) to my earlier book with Cambridge University Press, *Can a Darwinian Be a Christian?* Many people told me that they enjoyed and agreed with that book, but that they could not see how any right-thinking person could be a Darwinian. This is my answer to those people. The days have passed when I could make a trilogy, and follow with a book about how one must be a Christian, but I hope nevertheless the reader will sense my respect for a religion that I do not share. I have been trying out the ideas in this book for many years, most recently in my Gifford Lectures given at Glasgow University in 2001; in my Herbert Jennings Lectures given at Baylor University in 2003; and in my Robert Grant Lecture given at University College, London, in 2005. I have provoked enough discussion to take comfort in something Charles Darwin once said, namely, that everyone likes a false hypothesis because it is so much fun refuting it. Some of this material has been privately printed, and I have felt free to reuse it.

A book is a bit like an iceberg. The author's name is on the cover, but ninety percent is below the surface, with the names of many others who deserve credit. First and foremost, I want to pay my respects to the late Terry Moore, my editor for many years at Cambridge University Press, who died of cancer at too early an age. It was his idea that I write this book, and it was he who helped me find the tone and topics. Wherever Terry is now, I hope he can take time off from singing hymns or shoveling coal to read what he set in motion.

I never write a book without three people at my shoulder, encouraging and criticizing. The first is David Hull, philosopher of science and longtime mentor, now retired from Northwestern University; the second is Robert J. Richards, philosopher and historian of biology at the University of Chicago; and the third is Edward O. Wilson, entomologist, sociobiologist, and brilliant popular writer of Harvard University. Our shared love of evolutionary ideas infuse this volume. Here at Florida State University, many friends and colleagues have been ready sources of information and help. These include Joseph McElrath, who fed me all sorts of useful items about Darwinism and literature; Dean Falk, whose brilliant work on the "hobbit" (*Homo floresiensis*) is in itself a complete argument for the importance of Darwinian ways of thinking; Zach Ernst in the Philosophy Department, whose thinking is so much more rigorous than mine; Joseph Travis, whose knowledge of evolution is equaled only by his fund of good stories about individual evolutionists; and John Kelsay, who has taken time from his work on the Islamic concept of war to discuss with me many of the theological and metaphysical issues that I cover in this book.

My various assistants have been invaluable. Special thanks go to Charlie Alt, who did so much of the legwork. Martin Young did the artwork. At Cambridge University Press, Beatrice Rehl took over after Terry's death and has been a constant encouragement, and once again Russell Hahn converted a manuscript mess into something that can be read with ease. At home, my family – Lizzie, the kids, the dogs, the cats, and most of all the ferrets – made sure I did not become totally obsessed with this project.

Finally, let me say what a joy it is to dedicate the book to David Castle, Stephen Haller, Jean Lachapelle, and Eduardo Wilner. They were my four doctoral students at the University of Guelph in my final decade of teaching there (1990–2000). They could not be four more different people, and I could not be more proud of each and every one.

Introduction

> If I were to give an award for the single best idea anyone has ever had, I'd give it to Darwin, ahead of Newton and Einstein and everyone else. In a single stroke, the idea of evolution by natural selection unifies the realm of life, meaning, and purpose with the realm of space and time, cause and effect, mechanism and physical law.
>
> D. C. Dennett (1995)

This is the judgment of the philosopher Daniel Dennett about the English naturalist Charles Robert Darwin, who discovered the theory of evolution through natural selection, published in 1859 in his *Origin of Species*, the work that provides the ongoing framework for evolutionary studies today. It is a judgment with which I concur. Many do not. Most obviously, there are the many American evangelical Christians who take the words of the Bible absolutely literally and who hence assert that the world and its denizens were created by God, miraculously, some six thousand years ago, in the space of a week. Recently, these "Creationists" have been joined by believers of a more sophisticated ilk, the so-called Intelligent Design theorists, who argue that no natural account of origins can be adequate, and hence that all histories must make space for special interventions by some form of thinking being. These are people at one end of the religious spectrum, yet even those toward the other end – those who argue that Genesis must be interpreted metaphorically and that God did create according to laws of evolution – tend nevertheless to suppose that blind laws need help, that they need special pushes, to create the wonderful world of life and to alleviate the harshness of the Darwinian process.

It is easy and natural for those of us of a more secular bent to smile somewhat smugly at what we take to be the insecurities and failings of the religious. We do not reject evolution, even for humans, and we want no interventions from outside nature. But interestingly and depressingly, when it comes to Darwinism – natural selection as the chief causal process behind all organisms – large numbers of people stand virtually back to back with the religious critics. It is well known now that many students of literature and (particularly) those drawn to cultural studies have little but contempt for (and, one suspects, fear of) almost everything to do with science and technology. Notorious is their claim that there is no objective truth and that everything – science particularly – is a social construction, an epiphenomenon of the society in which it is produced. A scientific theory tells us no more about reality than does a political manifesto or a preacher's sermon. In this gloomy assessment, Darwinian evolutionary theory always has a special place – right at the center of the pit of damnation reserved for the very worst sinners. Supposedly, Darwinism reflects and justifies the grossest sins in our society – domination, greed, selfishness, sexism, and more. Those who have tried to portray Darwinism positively in literature must be tarred with the same brush. In the opinion of the many historians who espouse this ideology of science, all of this is hardly a surprise. Apart from the fraud and plagiarism that would make the most hardened internet-essay buyer blush, the coming of Darwinism was more an excuse for various people in the nineteenth century to park their prejudices in a respectable place, and this is a practice that has continued down to the present.

Many members of the social sciences – particularly in areas like sociology and cultural anthropology – feel much the same way. The very suggestion that humans might be animals, reflecting their biology like other brutes, is enough to bring on apoplexy in the nicest and best-qualified assistant professors. If not that, then firmly shut mouths and minds as tenure decisions loom. Darwinism supposedly leads straight to "genetic determinism" and other horrible philosophies that still have the odor of the vile world systems of the first part of the twentieth century. At the other end of the scientific spectrum, we now have physicists who tell us that their science can do it all for us – that there is no real need of natural selection. The laws of physics, unaided, can produce and explain everything worth knowing about organisms. Even the new areas of inquiry, like cognitive science, are getting into the act. People whose only knowledge of the living world has been filtered through the glow of the computer screen calmly tell us that their algorithms prove that the ideas of the *Origin* are now as outmoded as phlogiston theory and Ptolemaic astronomy.

What is truly surprising is that this skepticism is to be found in the biological sciences, even in evolutionary studies. For the past three decades, there have been well-qualified and articulate evolutionary biologists who have been showing so visceral a hatred of Darwinian thinking that one suspects that their objections cannot be grounded purely in theory or evidence. Tempting as it is to indulge in psychoanalytic hypotheses about lack of self-worth, the reasons are probably more prosaic. The more noted critics are avowed Marxists – one book dedication runs "To Friedrich Engels, who got it wrong a lot of the time but got it right where it counted" – and this bastard offspring of Hegelian idealism, itself an attempt to rejuvenate the traditional religion of the centuries, is taken to be reason enough to abjure any theory that suggests that biology might make humankind at one with nature. The underlying themes of Christianity run deep and emerge in unexpected places. What turns all of this from farce to tragedy – especially for me, a professional philosopher – is that these critics, the evolutionists particularly, have infected my own field of study. For more than a quarter-century now, there has been an apparently limitless flow of philosophical invective directed against Darwinism. Even when it is temperate, it is almost uniformly negative or belittling.

It is the nature of philosophy that its practitioners are drawn to attack received or standard positions – that is how we make our living, and there is no shame in that. But when you have a dominant scientific theory that its practitioners think is working magnificently well, and when one philosopher after another after another devotes large chunks of a career to proving that it is conceptually flawed and morally dubious, then one does start to wonder. It is all too reminiscent of those clever Jesuits who hauled Galileo over the coals for being a Copernican. Could it be that these most secular thinkers are a little scared that we humans might not be all that special after all? Can they not handle the awful truth that, after Darwin, the starting point of philosophical inquiry must be that we are the end product of a long, slow, natural process of selection rather than the creation of a good God miraculously on the Sixth Day? In the immortal words of Margaret Thatcher to George H. W. Bush when he showed signs of doubt and hesitation, when push comes to shove, are they getting a little wobbly? Thank goodness, there is yet time for repentance of sinners. Having spent a lifetime rejecting God and promoting Darwinism, the philosopher Anthony Flew has had a road-to-Damascus experience. Science cannot do it all. There must be something out there bigger than the both of us. "As people have certainly been influenced by me, I want to try and correct the enormous damage I may have done" (Wavell and Iredale 2004, 7).

At the risk of damning myself in the eyes both of sound scholarship and of God, let me be categorical. All of the critics of Darwinism are deeply mistaken. Charles Darwin was a good scientist, the biological revolution of the nineteenth century led to genuine understanding, and today's version of the theory is good quality science. It tells you important things about the real world. The integrity of evolutionists in general, and of Darwinians in particular, does not give great cause for complaint or alarm. It is of fundamental importance to philosophy to recognize the implications of Darwin's work for the major questions of knowledge (epistemology) and morality (ethics). Life would be much poorer without creative writers, and one welcomes the fact that they turn their attention to evolution. Whether what they say about Darwinism is a cause for concern is another matter. Finally, although, like all good science, Darwinism challenges religion, Christianity specifically, it can and should provide a positive and creative stimulus for religious people to think about their faith and move forward in a richer and deeper way. But let me not spend time telling you about what I believe. Let me turn at once to telling you why I believe what I believe.

CHAPTER ONE

Charles Darwin and His Revolution

It is not to be expected of Darwin that he should have been troubled by thoughts of fallibility, relativity, or indeterminacy; but only that he should have observed the standards of his own time. And it was by those standards that he was in arrears. Nineteenth-century science was sufficiently aware of the desirability of precision and standardization to make Darwin's tool chest seem distinctly unprofessional. In this, as in other respects, he gives the appearance of an amateur, an amateur even for his own day.

G. Himmelfarb (1959)

The above quotation is about Darwin as a practicing scientist and about how he went about things on a daily basis. It is typical of a certain strain of thought. At least here Darwin is only being labeled second-rate. His moral integrity is not being impugned. Others feel less constrained. Often we learn that Darwin's supposed virtues were equally his vices: "a conservative outlook in every respect except the evolutionary hypothesis; a failure to recognize or to relate his own ideas, his larger ideas, with those of others working in the same field; and a flexible strategy which is not to be reconciled with even average intellectual integrity" (Darlington 1959, 60). No wonder the conclusion is that "Darwin was slippery."

These are judgments made nearly fifty years ago, and in the time since then a veritable Darwin industry has grown up, publishing hitherto-unseen documents, assessing and reassessing the personality and actions of Darwin and his fellows, and looking at broader issues (Ruse 1996a). But in some respects things are little better. One prominent scholar has made a whole career out of arguing that what we truly had was a

"Non-Darwinian Revolution" (Bowler 1988, 2005). Others doubt that there was a revolution at all, and certainly do not think that it was truly Darwinian (Hodge 2005). It is argued that picking out one person, Charles Darwin, rather than any other is a matter of hero making after the event, and probably tells us more about our needs and interests than about anything that actually went on at the time (Secord 2000). There are others who agree that there was a revolution and that Darwin rode the crest of it, but (much as is suggested in the quotation above) who think that really Darwin was drawing on the efforts of others. The spade work was done for him (Herbert 2005; Lennox 2005). One variant of this line of approach is to suggest that although it is true that Darwin himself was English, it is a mistake to think that the Darwinian revolution was essentially English. Depending on the person writing, it is argued that the really important moves were made on the continent; France and Germany most commonly favored (Richards 2003; Corsi 2005).

So what did happen and who does deserve the credit? Let us see.

The Problem of Final Causes

The earliest evolutionists, as we would understand them – thinkers proposing the idea or fact of evolution, namely, common origins for organisms, and believing that everything occurs according to normal laws of nature – were people like the general man of letters, the Frenchman Denis Diderot, writing in the middle of the eighteenth century. We must ask two questions. First, why did it take so long for such an idea to emerge? Second, why did it emerge exactly when it did? You might think that answers to both of these questions rest ultimately on empirical discoveries, and to a certain extent you would be right. Although the major focus of the Scientific Revolution that had begun two centuries earlier was on physics and later on chemistry, increasingly attention was directed to the life sciences broadly construed. The invention of the microscope stimulated interest in the world that exists beneath our vision; journeys of exploration brought back fabulous new finds of various life forms around the world; and this is not to mention the side effects of technology and the growing methods of industry – the fossil finds, for instance, thrown up by mining and by the labors involved in building roads and (increasingly) canals. But none of these findings really speaks right to our questions. Back at the time of the ancient Greeks, four or five centuries before Jesus, there were speculations of at least a quasi- or proto-evolutionary kind. Empedocles suggested that there were disembodied parts of organisms floating around, and that sometimes these came together, and that if they worked, they cohered and reproduced. Why did such an idea not take off back then? There was one major reason why such an idea did not convince. The serious thinkers – people like the philosophers Plato and Aristotle – simply could not see how the intricate functioning parts of organisms, what we now call "adaptations," integrated into full living beings, could come about through blind, undirected law. The engineers call it Murphy's Law: if things can go wrong, they will. Blind law leads to disorder and to mess, not to complex entities working toward ends (Ruse 2003).

What then does lead to complexity working toward ends? The Greek philosophers had the answer: intelligence. Aristotle referred to the causes that bring on adaptive functioning as *final causes* (as opposed to things like efficient causes, which start things going), and for two thousand years it was these that made ideas of evolution simply not plausible. It was not old-fashioned prejudice but common sense. Where and what is the intelligence lying behind final causes? With the coming of Christianity, the great theologians – Saint Augustine and Saint Thomas Aquinas – argued that final causes point to the divine intelligence at the center of their religion. This yielded the so called Argument from Design or Teleological Argument for the existence of God. "We see that things that lack intelligence, such as natural bodies, act for an end, and this is evident from their acting always, or nearly always, in the same way, so as to obtain the best result. Hence it is plain that not fortuitously, but designedly do they [things of this world] achieve their end." From this premise, we move to the Creator behind things. "Now whatever lacks knowledge cannot move towards an end, unless it be directed by some being endowed with knowledge and intelligence; as the arrow is shot to its mark by the archer. Therefore some intelligent being exists by which all natural things are directed to their end; and this being we call God" (Aquinas 1952, 26–7).

In the Scientific Revolution, the notion of final causes came under heavy attack in the physical sciences. They were judged scientifically useless and misleading. The philosopher Francis Bacon referred to them as akin to the Vestal Virgins – decorative but sterile. Yet in the biological areas of science, it was agreed that it is impossible to study nature without making reference to ends, to intentions, to values. "For there are some things in nature so curiously contrived, and so exquisitely fitted for certain operations and uses, that it seems little less than blindness in him, that acknowledges, with the Cartesians [the followers of the seventeenth-century French philosopher and mathematician René Descartes], a most wise Author of things, not to conclude, that, though they may have been designed for other (and perhaps higher) uses, yet they were designed for this use" (Boyle 1688, 397–8).

This brings us to the eighteenth century, with people seeing that organisms need final causes for their explanation, and recognizing also that final causes seem inexplicable in purely natural terms. Evolution is a naturalistic explanation. Hence, evolution seemed an unreasonable position. Why then did evolutionary ideas start to emerge? Almost paradoxically, the reason lies in the Christian religion. As people started to find Christianity less and less compelling – as philosophers showed that it was unreasonable, as travelers brought back tales of other religions and other civilizations, as the move to an industrial world made the social force of the old beliefs less compelling and pertinent – they nevertheless sought alternatives in terms that Christianity had set. Although the Greeks had histories, they did not have a world history in our sense. They thought of the universe as eternal – going in endless cycles, with some limited variations here on Earth, but ultimately with no real direction. There was no creation out of nothing. Moreover, we humans were not the central focus of the action, however important we may seem to ourselves. Aristotle's God spent his time contemplating his own perfection and had no interest in us. The Jewish story of origins, taken up into Christianity, changed all of that. We have a beginning, a middle, and talk of an end. We have the creation of life from nothing. We humans have a special status, because we are made in the image of God. The world does not necessarily exist just for us, but we are the star players. We have the story of our Fall, but then comes the drama of Jesus and his sacrifice on the Cross, creating the possibility of our salvation. We have our roles to play, worshiping God and loving our neighbors. And finally, if everything works out, assuming that God is on our side and we have done what we should, we have the promise of eternal life.

People were looking for a non-Christian alternative, but set in the Christian terms – history, meaning, humans. Evolution told just such a story, offering rival answers to these same questions. It tells us where organisms came from – they started as primitive blobs way back when, and then grew and developed up to the forms that we have about us now.

It puts us humans up at the front of the picture, as the most important organisms, that to which all has been pointing. It gives us tasks to do, namely, to keep things going and to make sure that things do not fall back – even better, to keep things moving forward. And finally, it offers hope of a brighter tomorrow, if not for us, then for our children and our children's children.

The reader today might question all of this – we ourselves will be questioning much before this book is finished. But for now, leave how you think that evolution should be interpreted, and go back nearly three hundred years to the way that the first evolutionists thought of the topic. Powerful at that time was a growing belief in the possibility and importance of progress – the idea that through our own efforts we humans ourselves can make life better and more efficient, in the realm of culture and technology. France particularly (but then spreading to other countries) was a home of such speculations. The idea's greatest enthusiast, Jean-Antoine-Nicholas Caritat, marquis de Condorcet, saw progress as something that started with the new discoveries in science and the arts and medicine and so forth, arguing that they lead up from error and poverty and inequality, to truth and understanding and universal harmony. It was easy for a man like this to move from progress in the social world to progress in the biological world – from the simplest to the most complex, from the least valuable to the most valuable, from the monad to the man, as people were wont to say (Ruse 1996b, 2005b).

This all tailored nicely with a notion (the so-called Chain of Being) going back to Aristotle that one can put all organisms in a line, from the simplest to the most complex – namely, us humans (Figure 1.1). Then, with a progressivist form of evolution postulated, the move was usually made in a circular fashion in justification back to the social world. Such were the speculations of Erasmus Darwin (1794–1796), grandfather of Charles, a physician writing in England toward the end of the eighteenth century. They were also the ideas – at the beginning of the nineteenth century – of the man who provided the fullest overall picture of upward natural change, the French (sometime aristocrat) Jean Baptiste de Lamarck (1809). He saw a "spontaneous generation" of lower life forms from mud and dirt, and then an upward progression through the Chain of Being, until we reach the human form. Admittedly, he thought that sometimes there are diversions and vagaries – generally brought on by the heritable effects of use and disuse, as the giraffe's neck gets ever-longer through

FIGURE 1.1. Ramon Lull's ladder of ascent and descent of the mind (1305).

stretching to reach leaves on the upper branches of trees (the mechanism traditionally known as "Lamarckism") – but, overall, nature shows the progress that Lamarck, as a good French radical, thought is evidenced in the best human societies. This does not mean that Diderot and Erasmus Darwin and Lamarck and other evolutionists (including, by the end of his long life, the German naturalist and poet Johann Wolfgang von Goethe) paid no attention to empirical facts. One aspect of the living world – noted incidentally by Aristotle – is that there exist similarities, isomorphisms, between the parts of organisms of very different species. These similarities, like those existing between the bones of the forelimbs of vertebrates, today known as "homologies," are indeed powerful evidence of common ancestry, with one initial form having been molded to various ends (Figure 1.2). But the facts were always secondary to the ideology, that of progress.

What then of final cause? The most important biologist of the early part of the nineteenth century, the French anatomist Georges Cuvier, loathed evolution. This was in part because he thought it empirically false. Cuvier was at the forefront of uncovering and interpreting the fossil record, and was much aware of the fact that there are many gaps in the record between kinds of different forms. But a much bigger part of his dislike stemmed from the fact that evolution ignored or belittled final causes – and Aristotle was a hero to Cuvier.

> Natural history nevertheless has a rational principle that is exclusive to it and which it employs with great advantage on many occasions; it is the *conditions of existence* or, popularly, *final causes*. As nothing may exist which does not include the conditions which made its existence possible, the different parts of each creature must be coordinated in such a way as to make possible the whole organism, not only in itself but in its relationship to those which surround it, and the analysis of these conditions often leads to general laws as well founded as those of calculation or experiment. (Cuvier 1817, 1, 6)

Cuvier's point was simply that blind law does not lead to intricate adaptations like the eye and the hand. These seem still to show purpose or intention in their design and creation, and evolution simply does not speak to this. End of argument. Or, at least, end of argument until the arrival of Charles Darwin, for it was to this very issue that he spoke.

Charles Robert Darwin

Charles Darwin (1809–1882) was a child of the moneyed, British middle class (Browne 1995, 2002). His father, Robert Darwin, eldest son of Erasmus, was a successful physician and also an important conduit between aristocrats looking for mortgages for their lands and industrialists looking for secure properties on which to lend their cash. His maternal grandfather

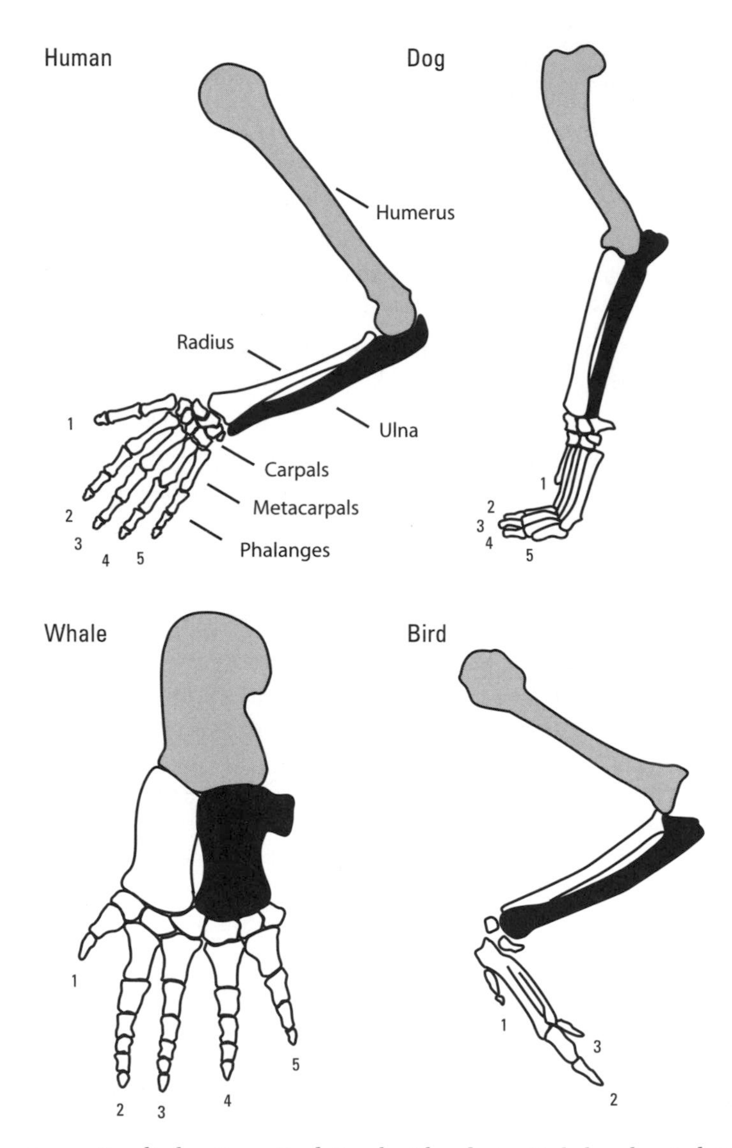

FIGURE 1.2. Forelimbs. From *Evolution* by Theodosius Dobzhansky et al. © 1977 by W. H. Freeman and Company. Used with permission.

was Josiah Wedgwood, the potter, one of Britain's most successful businessmen. More cash came young Charles's way when he married a first cousin, Emma Wedgwood, the daughter of Josiah's oldest son. It is worth stressing this fact, because we should not expect to find Charles Darwin kicking against the pricks, repudiating his comfortable background. He was not the Christian God making things from nothing. Inasmuch as Darwin was a revolutionary, it would be because he took things he received and rearranged them in an altogether new pattern (Ruse 1999a).

Charles Darwin naturally heard about the idea of evolution at an early age. He read his grandfather's works when young, and then after finishing school was packed off to Edinburgh to train as a physician. There too he met and became friendly with people with evolutionary inclinations. But there was no immediate conversion. Darwin decided to change his career choice. He went off to Cambridge University to train as an Anglican clergyman, and apparently had a very conservative view of life history. Things started to change in 1831, when Darwin was invited to sail aboard a British warship, HMS *Beagle*, as it charted the coast of South America, and it was the five years spent going around the globe that was to prove decisive. First, Darwin began to break from his Christian faith. He never became an atheist, and only in the final years of his life did he become an agnostic. But during the *Beagle* voyage he started to move away from Christianity – a theistic religion that places heavy emphasis on God's miraculous intervention in the creation – to what is usually called "deism," in which God is presumed to have created the world and then left everything else to the working out of laws. Influential here was the great geological work of the Scottish-born lawyer Charles Lyell. Darwin took with him the first volume of Lyell's *Principles of Geology* – a work that argued that, given enough time, unbroken laws can move mountains and create seas – and the later volumes were sent out as they appeared. Darwin became an ardent Lyellian, and he too wanted to explain phenomena naturally rather than through miracle.

A number of things shook Darwin's confidence in the stability of species, but the key insight seems to have been realizing the peculiar distributions of the reptiles and birds of the Galapagos Archipelago in the middle of the Pacific (Figure 1.3). Why should you have forms similar to but not identical to the mainland forms, and even more, why should you have different forms on islands literally within sight of each other? Darwin knew that on the main South American continent sometimes the same kinds could be found along the full length of the continent, from the steamy jungles of Brazil to the snowy deserts of Patagonia. The crucial conceptual leap was not made until Darwin returned home to England and was convinced that the island forms are genuine species. Then, early

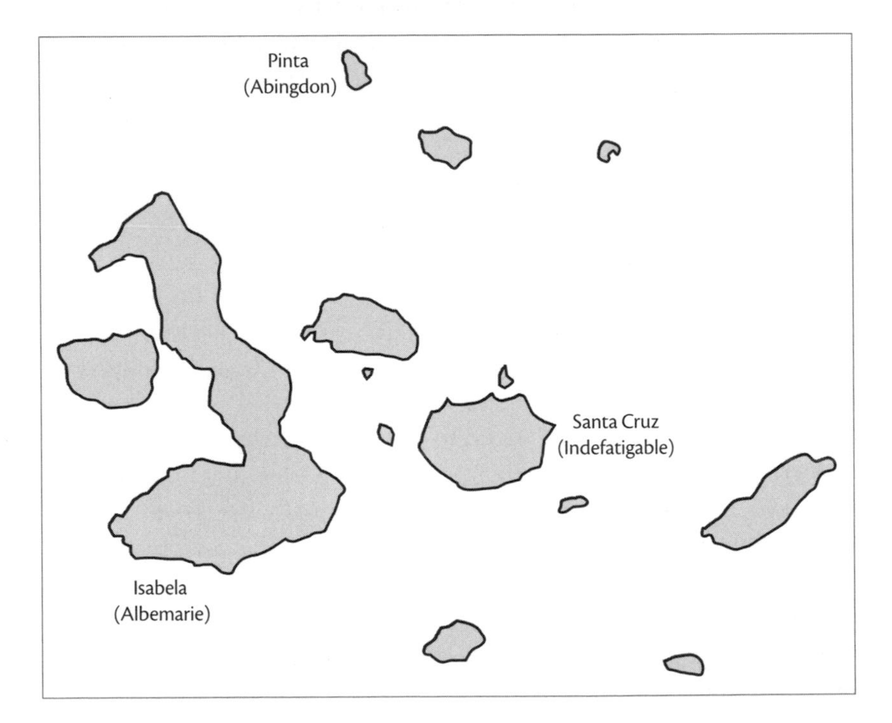

FIGURE 1.3. Galapagos tortoises. From *Evolution* by Theodosius Dobzhansky et al. © 1977 by W. H. Freeman and Company. Used with permission.

in 1837, he decided that there is only one solution – evolution, or what he generally called "descent with modification." It had to be that the early representatives of the reptiles and birds had come to the islands, and once there had spread and diversified. There is a real tree of life.

Darwin was a graduate of the University of Cambridge and realized – in the tradition of the great Isaac Newton – that one must find a cause, preferably a force-like cause. For eighteen months he worked hard (and secretly). He was much impressed by the success of animal and plant breeders and realized that their main tool of change is selection – one picks and breeds from the desirable organisms and rejects the others. But, until at the end of September 1838, Darwin could not see how to apply artificial selection to nature. Then he read a conservative political tract by the Reverend Robert Malthus, who argued that state support of the poor only exacerbates problems, for population numbers always outstrip potential food supplies. Given the inevitability of a "struggle for existence," Darwin moved straightaway to his mechanism of natural selection. Let me give the argumentation that he presented some twenty-one years later – after the naturalist Alfred Russel Wallace had sent Darwin an essay containing the idea of natural selection and sparked Darwin into writing and publishing his *Origin of Species*. First, Darwin argued to the struggle for existence.

> A struggle for existence inevitably follows from the high rate at which all organic beings tend to increase. Every being, which during its natural lifetime produces several eggs or seeds, must suffer destruction during some period of its life, and during some season or occasional year, otherwise, on the principle of geometrical increase, its numbers would quickly become so inordinately great that no country could support the product. Hence, as more individuals are produced than can possibly survive, there must in every case be a struggle for existence, either one individual with another of the same species, or with the individuals of distinct species, or with the physical conditions of life. It is the doctrine of Malthus applied with manifold force to the whole animal and vegetable kingdoms; for in this case there can be no artificial increase of food, and no prudential restraint from marriage. (Darwin 1859, 63)

Note that, even more than a struggle for existence, Darwin needed a struggle for reproduction. It is no good having the physique of Tarzan if you have the sexual desires of a philosopher. But with the struggle understood in this sort of way, given naturally occurring variation, natural selection follows at once.

Let it be borne in mind in what an endless number of strange peculiarities our domestic productions, and, in a lesser degree, those under nature, vary; and how strong the hereditary tendency is. Under domestication, it may be truly said that the whole organization becomes in some degree plastic. Let it be borne in mind how infinitely complex and close-fitting are the mutual relations of all organic beings to each other and to their physical conditions of life. Can it, then, be thought improbable, seeing that variations useful to man have undoubtedly occurred, that other variations useful in some way to each being in the great and complex battle of life, should sometimes occur in the course of thousands of generations? If such do occur, can we doubt (remembering that many more individuals are born than can possibly survive) that individuals having any advantage, however slight, over others, would have the best chance of surviving and of procreating their kind? On the other hand, we may feel sure that any variation in the least degree injurious would be rigidly destroyed. This preservation of favourable variations and the rejection of injurious variations, I call Natural Selection. (80–1)

Note that natural selection speaks to the issue of final cause, something that Darwin took very seriously not only because he knew of the arguments of people like Cuvier but also because at Cambridge he had been fully exposed to the Argument from Design (mainly through the writings of another Anglican clergyman, Archdeacon William Paley). Those organisms that are selected are those with favorable features, and given enough time, this leads to the evolution of things like the hand and the eye.

Finding a cause is important, but this is only the beginning. In the *Origin*, having introduced selection, and having given some brief remarks about variation and its possible causes, Darwin moved right on to apply his mechanism throughout the living world, taking in turn instinct, paleontology, biogeography, and then a mixed bag of areas including anatomy, systematics, embryology, and more. Finally, after a brief reference to the applicability of his theory to our own species, Darwin was ready for his famous conclusion.

It is interesting to contemplate an entangled bank, clothed with many plants of many kinds, with birds singing on the bushes, with various insects flitting about, and with worms crawling through the damp earth, and to reflect that these elaborately constructed forms, so different from each other, and dependent upon each other in so complex a manner, have all been produced by laws acting around us. These laws, taken in the largest sense, being Growth with Reproduction; Inheritance which is almost implied by reproduction; Variability from the indirect and direct action of the external conditions of life, and from use and disuse; a Ratio of Increase so high as to lead to a Struggle for Life, and as a consequence

> Natural Selection, entailing Divergence of Character and the Extinction of less-improved forms. Thus, from the war of nature, from famine and death, the most exalted object we are capable of conceiving, namely, the production of the higher animals, directly follows. There is grandeur in this view of life, with its several powers, having been originally breathed into a few forms or into one; and that, whilst this planet has gone cycling on according to the fixed law of gravity, from so simple a beginning endless forms most beautiful and most wonderful have been and are being, evolved. (489–90)

Darwin's other main contribution to the evolutionary question came some twelve years later, when he turned his attention directly to our species. In his *Descent of Man*, Darwin went beyond natural selection, and turned to a secondary mechanism – sexual selection. As it happens, this was an idea that Darwin had had from the first – organisms compete within the species for mates, and thus adaptations like the peacock's tail are produced. But in the 1860s, Wallace became enthused about spiritualism and argued that many human features – our large brains, our hairless skin, and the like – could not have been produced by natural means. Darwin agreed that these features may not have been produced by natural selection, but insisted that they must have been caused by natural means. Hence the role of sexual selection. Much of *Descent of Man* ranges through the full scope of the animal world, but Darwin never doubted that we humans are part of that world. Organisms compete for mates, and Darwin argued that it is this that was a major factor in human evolution. We belong to the overall picture.

The Darwinians

What happened after the *Origin* was published? Two things, mainly (Hull 1973). First, people overwhelmingly accepted the idea of evolution – the fact of evolution, that is. There were exceptions, and some never became evolutionists – some today do not accept evolution. But in Britain, Europe, and America, evolution became more and more acceptable, and often very rapidly. Second, people overwhelmingly rejected the putative Darwinian mechanism for evolution – natural selection. Again there were exceptions. In particular, those working on fast-breeding organisms like butterflies and moths were drawn to selection as an explanatory mechanism. We shall meet some of these people in later chapters. And no one said that natural selection could do absolutely nothing at all. But generally, selection was rejected as inadequate, and other causes were sought – some opted for natural causes, and others wanted at least some kind of divine intervention. Why was there this rejection? Often the issue was adaptation (final cause) or its supposed absence. For some, like Darwin's great American supporter, Asa Gray (1876), natural selection was inadequate to explain adaptation. Gray wanted to supplement things with divinely directed variations. For others, like Darwin's great British supporter, Thomas Henry Huxley (1893), natural selection was unneeded because adaptation had been overrated. Homology – explicable by evolution – was what really counted. But there were also scientific reasons why selection was not favored. Most significantly, Darwin had no adequate theory of inheritance and assumed (incorrectly) that most features blend in each generation. Critics pointed out that however strong selection might be, it could never overcome the effects of blending. The most significant new variations would be diluted beyond value before natural selection could do anything really effective.

In the light of this and other criticisms, selection was given but a minor role. Some opted for a kind of refurbished Lamarckism – the inheritance of acquired characteristics. Others went for jumps or saltations – what we today would call "macromutations" – taking organisms from one species to another. For those who thought adaptation important, these saltations would have been directed. For those who thought adaptation overrated, these saltations would have been random. Yet others favored a kind of evolutionary momentum, taking organisms along certain fixed lines – perhaps starting in the adaptive zone but eventually going beyond and against functioning adaptation. The baroque nasal appendages of the titanotheres were a favorite example of such a process, often called "orthogenesis" (Figure 1.4). Yet in a way, all of these alternatives strike one as being something of a red herring. It would have been possible to do selection-powered causal studies for all of the problems, had people wanted to – which is more than could be said for most if not all of the alternatives. The fact of the matter is that by and large no one was that much interested in a fully developed, causal theory of evolution.

Why? Primarily because (whatever had been the hope of Darwin himself) most people did not want to use evolution as (certainly not primarily as) a straight scientific theory (Ruse 2005b). They were far more interested in exploiting its potential as a kind of alternative to religion – what one might even go so far as to call a secular religion. They wanted to have

FIGURE 1.4. The titanothere. From H. F. Osborn, *The Titanotheres of Ancient Wyoming, Dakota and Nebraska* (Washington, D.C.: U.S. Geological Survey Monograph no. 55, 1929). Courtesy of the U.S. Geological Survey.

something that they could use to combat the dominant Christianity of the day – a Christianity that was (properly) seen as in league with the conservative forces and institutions of Victorian Britain (and similarly in other countries), and that would be challenged by the new progressive scientific attitudes and theories and technological implications. Evolution – a thoroughly progress-impregnated evolution – fit the bill. Although much of this was done in Darwin's name, in truth a far greater influence in the post-*Origin* period was Darwin's fellow Englishman Herbert Spencer (Richards 1987). An ardent Lamarckian (for all that he discovered the idea of natural selection independently), Spencer was unambiguous in seeing organic evolution as part and parcel of the overall upward progress that characterizes literally everything: as a move from the undifferentiated to the differentiated, or as what he called a move from the homogeneous to the heterogeneous:

> Now we propose in the first place to show, that this law of organic progress is the law of all progress. Whether it be in the development of the Earth, in the development of Life upon its surface, in the development of Society, of Government, of Manufactures, of Commerce, of Language, Literature, Science, Art, this same evolution of the simple into the complex, through successive differentiations,

holds throughout. From the earliest traceable cosmical changes down to the latest results of civilization, we shall find that the transformation of the homogeneous into the heterogeneous, is that in which Progress essentially consists. (Spencer 1857, 2–3)

Everything obeys this law. Humans are more complex or heterogeneous than other animals, Europeans more complex or heterogeneous than savages, and the English language more complex or heterogeneous than the tongues of other peoples.

When did this Darwinizing – as we might call it – come to an end? Turn the question around somewhat, and leave until a later chapter a more detailed discussion of progress and how that notion has fared in evolutionary circles. Ask rather when it was that evolution finally took causal questions seriously and when it was that Darwinian selection was seen to be the key causal mechanism. The crucial event was the discovery of the fundamental principles of heredity, called "Mendelian genetics" after the Moravian monk who had discovered them in the second half of the nineteenth century (Provine 1971) (Figure 1.5). At first, such principles – which supposed very stable units of heredity and function that were transmitted unchanged from generation to generation, a process broken only by the occasional nondirected random change (mutation) – were thought antithetical to Darwinism, contradictory. But soon it was seen that Darwinian natural selection and Mendelian particulate heredity are complements, the one filling in gaps and answering questions raised by the other. Around 1930, three people – Ronald Fisher and J. B. S. Haldane in Britain and Sewall Wright in America – put together these ideas in systematic form, and thus came the rise of so-called population genetics, the basis of a refurbished Darwinism.

Using the theoretical skeletons provided by the population geneticists, the experimentalists and fieldworkers moved in to add empirical flesh (Ruse 1996b). In Britain, the most important was the Oxford-based E. B. ("Henry") Ford, who founded what he called the school of ecological genetics. Ford worked with Fisher and gathered around himself a group of dedicated naturalists who went out into the field studying butterflies and snails and other organisms. Particularly noteworthy was the work of A. J. Cain and Philip Sheppard on snails and their adaptive markings. A master at networking and inspiring, Ford found funds for what became a thriving enterprise, in particular persuading the Nuffield Foundation

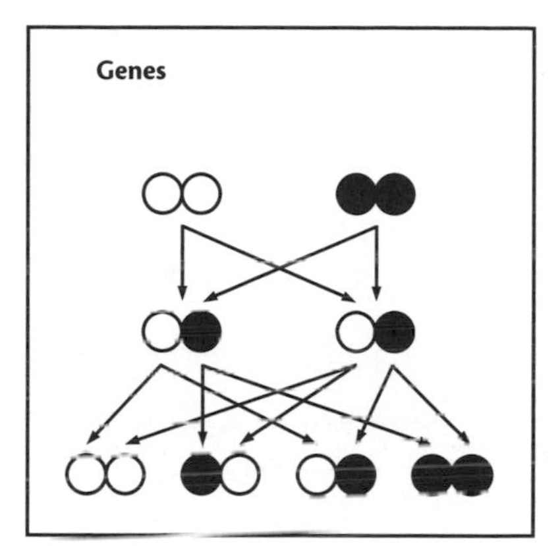

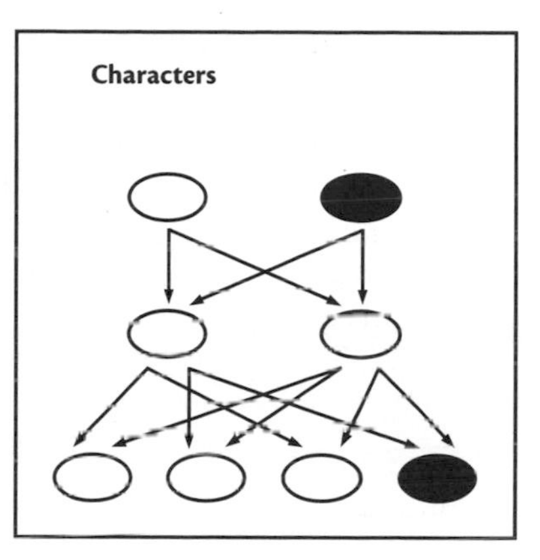

FIGURE 1.5. Mendel's first law (also known as the law of segregation). Mendelian genetics supposes that there are two factors, "genes," that affect or cause any given physical character. Sometimes, as in the illustration above, one gene (the "dominant" one) masks the effects of the other gene (the "recessive" one). Genes are passed on from one generation to the next entire, unless there is a spontaneous change ("mutation"). Each parent contributes one gene to each offspring, and it is pure chance (within the theory) as to which of the parental genes is transmitted in any instance. The genes are carried on string-like entities ("chromosomes") within each cell of the body; the place that a particular gene occupies on the chromosome is called the "locus." The various genes that can occupy the same locus of individuals in a population are known as "alleles." Overall, the collection of genes in an organism is called the "genotype," and the physical body is the "phenotype." If both alleles at a locus in an organism are the same, the organism is said to be "homozygous" with respect to that locus; if they are different, then it is "heterozygous." Mendel's second law (also known as the law of independent assortment) states that the transmission of genes at any one locus is independent of the transmission of genes at any other locus. In fact, this is true only of genes on different chromosomes. Figure 2.3 (p. 23) from *Evolving: The Theory and Processes of Organic Evolution* by Francisco J. Ayala and James W. Valentine. Copyright © 1979 by The Benjamin/Cummings Publishing Company, Inc. Reprinted by permission of Pearson Education, Inc.

that the work he and his fellows were doing in England's hedgerows and woods and meadows had strong implications for the human world, showing how diseases can spread and be maintained and fashioned by natural selection. (See Ford 1964 for his overview of the school's achievements.)

In America, the leading figure in building the new Darwinian theory – something that came to be known as "neo-Darwinism" (in Britain) and the "synthetic theory" (in America) – was the Russian-born Theodosius Dobzhansky. In 1937, he produced his masterwork, *Genetics and the Origin of Species*, and with this founded a whole school dedicated to studying the genetic changes of populations, in nature and in the laboratory. In short order, there came *Systematics and the Origin of Species* (1942) from the systematist-ornithologist Ernst Mayr, *Tempo and Mode in Evolution* (1944) from the paleontologist George Gaylord Simpson, and then *Variation and Evolution in Plants* (1950) from the botanist G. Ledyard Stebbins.

Summing Up

This brings us to 1959, the hundredth anniversary of the *Origin*, an event much celebrated as the triumph of Darwinism (Smocovitis 1999). Let us leave things here. Our question now is where all of this leaves Charles Darwin and the Darwinian revolution. In one way, the rest of this book is going to try to tell you that, and a full evaluation cannot be given until much later. For a start, we need to look at the scientific work since 1959 and see if it is still Darwinian, or if the people back then were mistaken in thinking that finally evolutionary studies were on the right track. But enough ground has been covered already to make three fairly solid points.

First, Charles Darwin himself. There is no doubt that he is for the historian a bit of a puzzle, and that there was about him something of the amateur. The critics are right. His hands-on work was pretty low-tech; his mathematical abilities were gamma minus; and his foreign language skills were about on a par with English cuisine. He never worked for a living; his support came from his family's wealth; he did not himself go out and found areas of scientific research in universities, and much more. What I find more striking than his crude ways of experimentation – after all, back then biology was often pretty primitive in these respects, with little more than a scalpel and a ruler – is the way in which so often, especially in the *Origin*, he writes for a general audience rather than just for specialists (which latter he was certainly capable of doing, as several long works on barnacle classification from the early 1850s well attest). The fact is that Darwin's patrons were people like his father and his uncle (and father-in-law), and he wrote with them in mind, even after they were long dead. Although Darwin desired the respect of professional scientists, he usually wrote for a wider audience.

Having said this, however, one must add that Darwin himself was solidly in the scientific group and seen as such (through the honors, like fellowship in the Royal Society). More than anything, he saw that the big problem for evolution was that of final cause, and he realized that any adequate solution had to incorporate a cause that spoke naturalistically to final cause. Darwin saw this, and he cracked the problem – the problem of two and a half millennia. Leaving for later the obvious questions and reservations, Darwin saw the big issue, and he answered it – not by chance, not by luck, but by hard work. And genius.

Was there a revolution, and was it Darwin's? Well, yes, there was a revolution. What was it? At a more limited level, it was moving to an evolutionary view of life's history from a static one – or from no history at all. One almost defines the meaning of the word "revolution" by this. One moves from a view of life created by a good God, a short while ago, in a very limited time and through miracles, to a slow, naturalistic (lawbound) view of origins, through a mechanism that has no intention, no goal. At a more general level, there was a move from a religiously inspired view of nature to one that does not necessarily have to be like this. To a view that is potentially materialistic, meaningless. Again, if this is not a revolution, I do not know what is. Obviously, how you assess Darwin hinges on how you think of the revolution. At the more limited level, he was not the first to think of evolution. He did come up with the cause. To use the obvious analogy, he was Newton to the Copernicus of Diderot and Grandfather Erasmus and Lamarck and the others. Darwin was certainly not a man alone, but he was very important, arguably the most important – especially when we look at the topic of the next chapter, providing evidence for the very fact of evolution.

At the more general level, one would dilute probably Darwin's significance more. The philosophers like David Hume and Immanuel Kant certainly deserve their acknowledgments for moving us to the new world (Ruse 2003). The same is true of those who wrote about religion and politics and much more. But even here, Darwin has a major – perhaps still the major – role, for it was he who showed how you can get final cause without a God. Even Hume had to admit that this problem defeated him. To use the words of the English popular-science writer Richard Dawkins (1986), only after Darwin was it possible to be "an intellectually fulfilled atheist." Whether Darwin's revolution was truly an English revolution was of course connected with all of those issues. It is certainly true that the ideas of many other nationalities fed in. It is also true that Darwin himself worked from English materials as he moved to natural selection – the success of the breeders of the agricultural world in which he grew up; the significance of Malthus's arguments about population control, a matter of much interest and concern in mid-Victorian England. Perhaps it is best just to leave things at that.

What about the charge that we privilege Darwin unfairly, neglecting others of importance? Well, the answer of course is that we do privilege Darwin, and we do so because of what came after. Darwin at the time did not accomplish what he set out to accomplish. He wanted a professionally based area of biological science, using natural selection as its main investigative tool. What he ended up getting was a kind of religion substitute, with progress at its heart, and with little interest in mechanisms. But later, in the twentieth century, his ideas were picked up and used for what they are worth, as a means to peer into the mysteries of nature. And this – subject to later discussion – is where we are today. But privileging is not necessarily wrong. Darwin is interesting and important because of what we today believe and do. No one today believes in the inheritance of acquired characteristics. That is why Lamarck is someone of much more limited interest, even to the French. You might complain that Darwin's fame and significance is riding on the backs of those in the twentieth century, like Fisher and Wright, who did make evolution into a fully integrated, selection-based theory. To an extent this is true, rather as Mendel's present fame rests on the backs of those who made his ideas into a full theory. But (much more so than Mendel) Darwin did point to the way in which such a theory might be formed, and were he alive today he would rightly think that his dreams had come true.

Hence, with all of the qualifications, we can be generous. There was a revolution; it was a Darwinian revolution; and Charles Darwin earned the status that he has now.

CHAPTER TWO

The Fact of Evolution

The believer in God, unlike her naturalistic counterpart, is free to look at the evidence for the Grand Evolutionary Scheme, and follow it where it leads, rejecting that scheme if the evidence is insufficient. She has a freedom not available to the naturalist. The latter accepts the Grand Evolutionary Scheme because from a naturalistic point of view this scheme is the only visible answer to the question *What is the explanation of the presence of all these marvelously multifarious forms of life?* The Christian, on the other hand, knows that creation is the Lord's; and she isn't blinkered by a priori dogmas as to how the Lord must have accomplished it. Perhaps it was by broadly evolutionary means, but then again perhaps not. At the moment, 'perhaps not' seems the better answer.

A. Plantinga (1998)

Thus writes Alvin Plantinga. He is no wild-eyed fundamentalist, not an evangelical Christian of the crudest type. He is a Calvinist, North America's most distinguished philosopher of religion, and a professor at the University of Notre Dame. There is much in his charge that needs unpacking and discussing. As we begin – more broadly, as we begin a systematic examination of the claims of Charles Darwin and of those who follow in his tradition – let us follow a customary and convenient practice of dividing the evolutionist's problems into three (Ayala 1985). First, there is establishing the very *fact* of evolution – proving that organisms came from other forms, perhaps just a few, perhaps even originally from nonorganic material, by a process or processes that are entirely natural, that is to say, governed by and only by unguided regularities or laws. Second, there is the question of the *path* of evolution. Is it tree-like, for example?

Does it show different patterns at different times? Can we put dates on events? Technically, this is known as a matter of "phylogeny," as opposed to the development of the individual, "ontogeny." And finally, there is the puzzle of *mechanisms* or *causes*. What makes things happen? Darwin's answer was natural selection. We have to see how adequate an answer this really was.

Fact or Theory?

Start with the first of these questions, the main focus of Plantinga's attack. Why should one accept the fact of evolution in the sense just explicated – gradual growth of forms from fairly primitive types, by a natural (that is, lawbound) process? Many, particularly in America today, join Plantinga in rejecting this idea. One of his points is that the debate about evolution is no mere matter of fact, but more a question of philosophy, namely, commitment to so-called naturalism. We shall have to consider this charge as well as the empirical evidence, but before we plunge in, let us clear up one confusion. One often hears the claim that "evolution is a theory, not a fact," meaning apparently that even if it did all that it claimed, it could only achieve some kind of second-class status. At best, supposedly, evolution belongs to that nether-world of dubious claims, like those about the identity of Jack the Ripper or alternative assassins of President Kennedy. It can never be really true, in the sense that George Washington truly kicked out the English and the North truly won the Civil War.

Simply – without judging the merits of evolution as such – this claim is based on confusion, pivoting on an ambiguity about the meaning of the notion of a "theory." Sometimes, one does mean by "theory" the same as "iffy hypothesis." If I say that I have a theory about the true identity of Jack the Ripper or about the Kennedy assassination, then you know that what I am going to propose is a hypothesis – a conjecture – that for the rest of us may or may not be true, but probably will never be decided definitively. But this is not the only sense of "theory." Sometimes one means "theory" as in "body of laws used to explain a range of phenomena." If I say that I accept Einstein's theory of relativity or the theory of continental drift through plate tectonics, it is in this second sense that I am using the term. It may well be that the theory is pretty solid – well confirmed – and that no reasonable person would doubt it. Continental drift is now surely in this category. It may well be that the theory is doubtful or almost certainly false. Phlogiston theory falls into this category. But this is all beside the point with respect to this use of the word "theory." "Theory" in this sense refers to the law network itself and not to its truth status.

In speaking of evolution, we should keep this point in mind. It may be that the fact of evolution is indeed a bit iffy – that it is only a theory in this sense. But logically it does not have to be, any more than Washington beating the English has to be. Evolution may be a well-confirmed fact – in which case, it is not a theory in this sense. It is certainly the case that evolution is a theory when it comes to mechanisms – as in the Darwinian theory of evolution. This usage tells you nothing in itself about whether Darwinism is well confirmed. It refers only to the fact that you have a law network with a mechanism. So it is certainly true that evolution is a "theory" in one (the second) sense, but it does not follow that it is necessarily a "theory" in the other (the first) sense. It is quite possible that evolution is a theory *and* a fact. Thus, to say that evolution is only a theory and not a fact may be true, but it is not necessarily true. This is something that must be shown, an inquiry to which we now turn.

Darwin's Three Approaches

Darwin may not have convinced people of natural selection. He did convince them of the fact of evolution. Therefore, let us turn to the *Origin* to see how it was done. Somewhat confusingly, Darwin does not separate out the argument for the fact of evolution from his argument for his mechanism of evolution, but focusing on the fact, we can see that there are three strategies. First, Darwin turns to the world of the animal and plant breeder to convince us that significant change can in fact occur, and that we ought to accept the analogy that significant change can and does occur in nature. Second, Darwin presents his mechanism of natural selection, with the implication that if a mechanism as strong and powerful as this is always operating, there has to be some effect, and this effect will be full-blown evolution. Third, Darwin goes through the whole gamut of biological areas of inquiry, showing that they are explained by evolution and, conversely, that they support evolution. Let us leave the second of these strategies for a later chapter – for a time when it did start to convince people – and take, in turn, the first and third. Obviously, our central concern is whether it is proper today to accept evolution as a fact. So, although we will start with Darwin and his strategies, we can feel free to move on to consider evidence that we now have at the beginning of the twenty-first century.

The Direct Evidence

What about the direct evidence, the evidence from observation, as it were? There are two parts to this: that which is done intentionally or inadvertently by humans, and that which occurs in nature (Jones 2001). With respect to human-produced evidence, Darwin himself instanced breeding both of animals and of plants, and the same is still strong evidence today. The domesticated dog, *Canis familiaris*, is the most obvious instance, coming in a huge number of breeds and varieties, from the Great Danes and other monsters down to the little lap dogs like the Mexican chihuahua, from the extremely hairy Afghan to a number of hairless breeds, from the elegant greyhound to the stubby Pekinese, from the social Labrador to the aggressive bull terrier (Serpell 1995). In the ten thousand or so years since the dog was domesticated, it has been molded and twisted into an incredible number of forms and shapes and personalities. In the plant world, much the same can be said of the corn plant, maize (*Zea mays*). It is the most remarkable case of human-caused change, from the wild grass teosinte (*Euchlaena mexicana*) to the tall, straight, strong plant that we see in the fields today, with its huge cob bursting with ripe kernels. So different do these two plants seem that they once were even put in different genera. But it turns out that they are in fact the same plant, separated by a number of gene changes. In the last five to ten thousand years in North. America, agricultural workers have changed it from a weed into one of the most desired of all plants (Mangelsdorf 1974).

What of non-human-produced evidence? Turn to nature for direct evidences of evolution. There are literally hundreds of documented examples of changes occurring, either for purely natural reasons or because humans have somehow altered the environment. Probably the most famous case of all concerns what is known as "industrial melanism" – as plants and buildings got dirtier in the nineteenth century, because of the Industrial Revolution, many organisms got darker in tandem, almost certainly because this gave them better camouflage than they otherwise would have had. Lepidopterists (students of butterflies and moths) particularly noticed the change, a direct function of the rapid reproductive rates of their subjects and hence the speed with which evolutionary change could occur and be spotted (Majerus 1998). One leading authority, at the end of the nineteenth century, was explicit:

> The speckled peppered moth as it rests on trunks in our southern is not at all conspicuous and looks like a . . . piece of lichen and this is its usual appearance and manner of protecting itself. But near our large towns where there are factories and where vast quantities of soot are day by day poured out from countless chimneys, fouling and polluting the atmosphere with noxious vapours and gases, this peppered moth has during the last fifty years undergone a remarkable change. The white has entirely disappeared, and the wings have become totally black. As the manufacturers have spread more and more, so the 'negro' form of the peppered moth has spread at the same time and in the same districts. (Tutt 1891)

We shall meet this example again, for the major work done on this phenomenon has recently been held up as an example of shoddy if not fraudulent science. For now, it is enough to stress that no one denies that real change has occurred. Moreover, insects were not the only organisms that the early post-Darwinians saw as changing. Birds were another item, and so were marine creatures. One of the most important of all early Darwinians (in the sense of someone who embraced natural selection as the chief cause of change) was the English biologist Raphael Weldon (1898). He studied crabs, first in the bay of Naples and then in the mouths of rivers in the South of England. He found that there were significant and ongoing changes in the dimensions of the crabs and that this was almost certainly linked to the amount of silt in the water – where river works changed the amount of silt, the crabs changed in parallel. Specifically, their frontal filters are reduced in size to prevent clogging. Weldon had some pretty strong opinions about how the change occurred, and we shall come back to these later.

In the twentieth century, perhaps the most famous case of change centers on disease resistance. It is notoriously well known that diseases rapidly evolve mechanisms to overcome or sidestep the various methods that humans have devised to protect themselves and their plants and animals from harm. Starting in 1914 with the first recorded example of resistance (the resistance of the San Jose scale to lime sulfur), over 500 different species of insect and mite have been recorded as having developed resistance to various insecticides. Most of these came after 1940, with the increasingly widespread use of DDT to combat malaria by killing off the mosquito. The reasons for the changes pose no mystery. Almost all populations of pest have one or two individuals that carry natural protections against the insecticide, and it is these that multiply and spread through the populations. Some of the natural protections actually protect against the insecticide, making it unable to attack the pest. Others enable the pest to digest and render harmless the insecticide. But the overall effects are the same – the pests develop natural protections against the insecticide, leading to a need for increased use of the insecticide or a desperate search for ever-new forms of agricultural protection (Committee on Strategies for the Management of Pesticide Resistant Pest Populations 1986).

At this point, both in the human-controlled world and in the world of nature, we must face a critic. Agree that we have many wonderfully transformed organisms – dogs, maize, moths, crabs, diseases. What of the objection that although one has these various forms, nevertheless they show – biologically speaking – only surface changes? What could one have in mind here? What would be – biologically speaking – a deep change? Most obviously, one would be thinking in terms of "species." It was after all the finding that the birds of the Galapagos truly were from different species that made Darwin into an evolutionist (although note that species is a concept that so far has been curiously absent from our discussion, particularly given that Darwin's masterwork was the *Origin of Species*). In fact, although Darwin had certainly thought seriously and deeply about these issues, his discussion on the topic is not very systematic (Beatty 1985). The issue, as he realized fully, is that the world of organisms varies over many different forms, and this variation is not continuous. Organisms form groups – that of course is the reason why classification is possible. Like everyone else (including us today), Darwin accepted the system of classification due to the eighteenth-century Swedish biologist Linnaeus, who saw that organisms could be assigned to nested sets – that is, several different groups at successively more comprehensive levels, until at the highest level there are only two or three groupings (Figure 2.1).

Like everyone else (including us today), Darwin also accepted that the groups at the lowest generally acknowledged level, species, seem in some sense to be more real or less arbitrary than groups at the higher levels, like genera or families. There seem to be real bonds between the members of species and gaps between species. What is the nature of the bonds and the consequent gaps? There were – and are – various proposals.

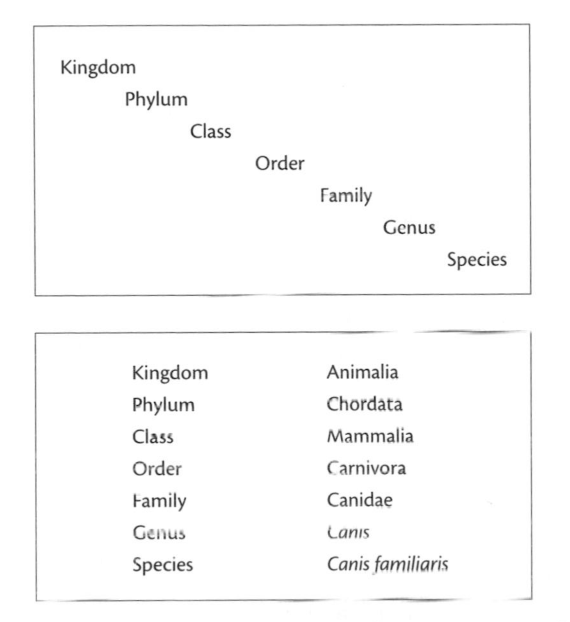

FIGURE 2.1. The Linnaean taxonomic system. Each organism – say, the dog Polly – is assigned a class (a "taxon") at each of seven levels ("categories") that get ever more inclusive as one rises from bottom to top. The lowest level is that of the species, which always includes the name of the taxon at the next level, that of genus. Hence, Polly is a member of the species *Canis familiaris*, of the genus *Canis*, and so forth. (Conventionally, specific names and generic names are italicized.) Humans, *Homo sapiens*, are (like dogs) animals, chordates, and mammals, but not carnivores. Hence, Polly and Charles Darwin do not share a taxon until the level of class. Obviously, the Linnaean system does not simply classify, but also shows relationships. Darwin argued that these relationships reflect descent. Refinements to the seven basic categories can be added. Divisions within a species (subspecies) are usually noted by adding a third name after the first two, as in the species of American fruitfly *Drosophila willistoni*, divided into the subspecies *D. w. willistoni* and *D. w. quechua*. (Where there is no ambiguity, after the first mention initials are often used.)

As a matter of empirical practice, one usually has to rely on some sense of similarity. Organisms in the same species look alike, and organisms in different species do not look alike. A cow looks like a cow and not a horse. But Darwin and others, then and now, realized that similarity is itself a bit of a subjective notion, and he and others tended to think more basic at some kind of causal level – at some kind of biological causal level – was the actuality or potentiality of reproduction. Species are collections of organisms with the members forming a breeding group – cows breed with cows – and reproductively isolated from the members of other groups – cows do not breed with horses.

Later we will turn to questions about the causes of species – Darwin's thinking and ours – but now, returning to the imaginary critic, what of the objection that thus far we have evaded questions about species and their origination, speciation? Perhaps he is a critic with good reason, for we have given no examples that cover the topic. All dogs are interfertile, and more than this, all can interbreed with other supposed separate species, notably the wolf, *Canis lupus*. Likewise in the vegetable world. You can cross maize with teosinte, and indeed it was this power that first enabled breeders to work out the close relationships. This objection is well taken, or at least it is more or less well taken. A poodle can certainly breed with a Labrador, as many pet owners have discovered to their dismay. But it is not quite true to say that no gaps have been opened up. The reasons for lack of interfertility can take many forms, physical as well as physiological. The largest dogs are a hundred times the size of the smallest dogs. Although the desire may be there, if one had only Great Danes and chihuahuas, it is hardly likely that one would have a totally interfertile group. The males of the smaller variety could not mount the females of the larger variety, and the females of the smaller variety carrying the offspring of males of the larger variety would probably die in the attempt to carry puppies well over the size for which they are fitted. Note also that, generally, domestic animals and plants have not been bred to be new species as such. A dog breeder takes infinite care to get a desired coat or stance or whatever. There is little or no interest in making the breed infertile with other breeds. This lack of interest in interfertility is perhaps changing now that genetically modified foods are becoming available – no one wants to give a weed an ability to resist herbicides – but this is a somewhat artificial situation that does not really apply here (Ruse and Castle 2002).

As a nonartificial situation in the plant world – for those for whom it is speciation or nothing – there is the apple maggot fly (*Rhagoletis pomonella*) (Feder, Chicote, and Bush 1988). This major pest, found all over the northeastern United States, also parasitizes the hawthorn tree, a close relative of the apple, and it turns out that the maggot fly (a North American native) started life on the hawthorn (also a North American native) and only later moved to the apple (introduced from Europe less than three hundred years ago, with the first maggot parasites noted in the nineteenth century). One might think that since apples and hawthorns all occur in the same ranges, the two sets of maggots – apple and hawthorn – would indeed be one, and that they would freely interbreed. It turns out, however, that the apple and hawthorn populations are quite different biochemically, and that although they can indeed interbreed, individuals from the two populations show significant preferences for their own respective hosts. Apple maggots prefer apples, and hawthorn maggots prefer hawthorns. The biochemical differences are not random but correlate significantly with breeding times, in particular making the apple maggots ready to breed a month earlier than the hawthorn maggots, something that obviously appeared in response to the fact that apples ripen (and hence are available to be parasitized) a month before hawthorns. Although one does not yet have full species, at the least one has species in the making.

As promised, there will be more on species and their formation in later chapters. For now, it is just not true that we have no direct evidence of the formation of such groups.

Consilience of Inductions

Stories of change can be told about the human realm, particularly in the realm of disease resistance – for instance, about why a single shot of penicillin used to be a cure for syphilis while small doses are ineffective today. The point is made. Change, human-caused or purely natural, does occur and has been observed – repeatedly. But obviously not change of the reptile-into-bird, hippopotamus-into-whale, monkey-into-human kind. In the short time span of human observation, one should not and does not expect this. If the direct evidence were all one had to go on, the case for evolution would be suggestive but unproven. This means one must turn to indirect evidence, and at once it is important to scotch one objection. Again, a version of the theory-not-a-fact objection raises its ugly head. How can one be sure on the basis of indirect evidence? It is bad enough that we have had to take on trust the claims about dogs and their domestication ten thousand years ago, but now the demand is that everything rest on indirect evidence. This can never be good enough.

We will tackle the specific case of evolution in a moment. At a general level, let me point out at once that not only do we use indirect evidence, we rely greatly on it in everyday life, and there are times when we would prefer it to direct evidence. In the case of sexual assault, which would you prefer – DNA evidence or eyewitness testimony? (Figures 2.2, 2.3) It is notorious

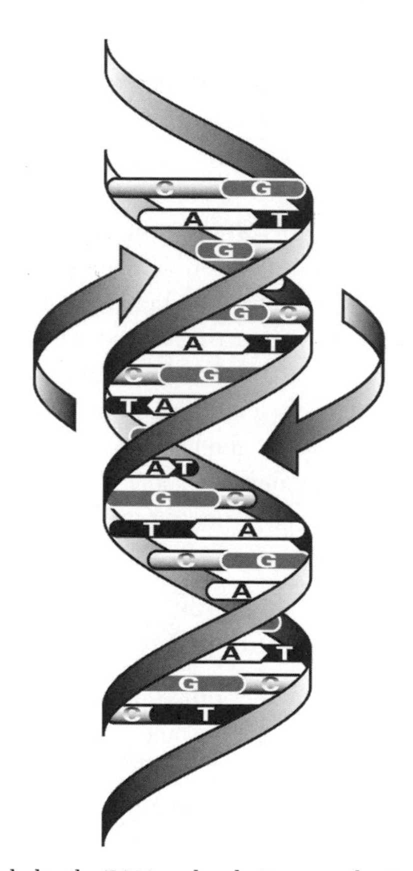

FIGURE 2.2. Double helix: the DNA molecule. In 1953, the American James Watson and the Englishman Francis Crick found that the deoxyribonucleic acid molecule (DNA) is a double helix, that is, two long molecules that twine around each other. This "molecular gene," the DNA molecule, carries information in the form of the ordering of four smaller molecules, or bases: adenine, A; cytosine, C; guanine, G; and thymine, T. Adenine always matches up with thymine, and cytosine with guanine.

that eyewitness testimony, especially under times of great stress, is highly unreliable. But DNA evidence is quite another matter, and convicts or exonerates on a regular basis. Of course, not any indirect evidence is good enough. There is indirect evidence and indirect evidence. To see what we really want, start with the plot and theme of a detective story by the English man of letters and Catholic apologist G. K. Chesterton (1986). His hero is a priest, Father Brown, and the particular problem being faced is set in a Scottish castle, whose owner (the earl of Ogilvy) has recently died and whose sole inhabitant is a near-mute and somewhat

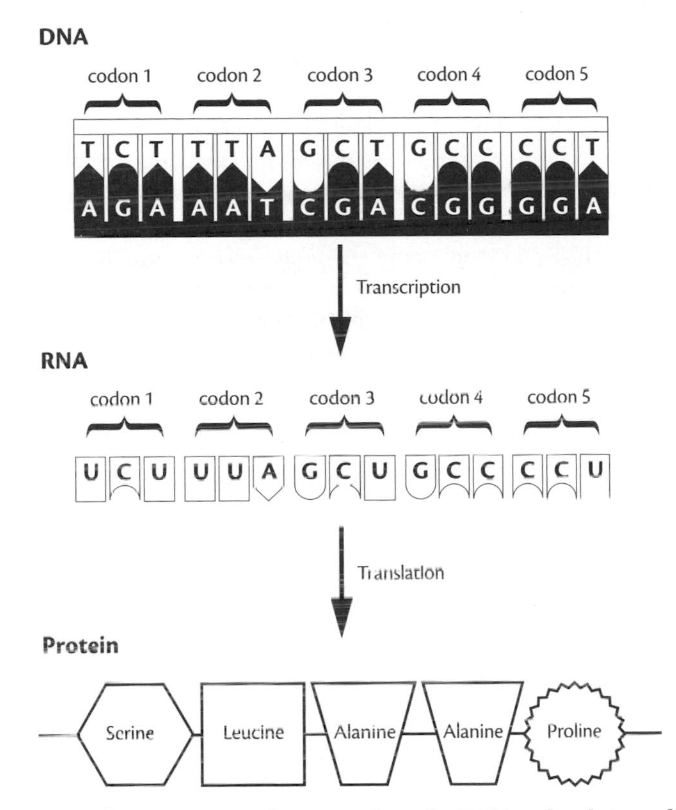

FIGURE 2.3. Making proteins. Information from the DNA molecule is read off (transcribed) by another macromolecule, ribonucleic acid (RNA), which is identical to DNA except that thymine is replaced by uracil, U. RNA has the job of translation, namely, picking up and joining the components (amino acids) of the building blocks of the cells (proteins). There are twenty different amino acids, and since there are only four bases, the minimum number of bases that could "code" for so many kinds is three. Since there are sixty-four different combinations that one can make with triplets (known as codons) from four bases, there is a lot of redundancy in the system, with different codons picking up the same amino acids. Adapted with permission from K. J. Niklas, *The Evolutionary Biology of Plants* (Chicago: University of Chicago Press, 1997).

mad retainer, Israel Gow. Summoned by a regular police officer, Father Brown asks that a candle be lit to dispel the gloom. Curiously, there are candles – twenty-five of them – but no candle sticks. And this is not all. The policeman reads aloud the list of oddities.

"First item. A very considerable hoard of precious stones, nearly all diamonds, and all of them loose, without any setting whatever. Of course, it is natural that the Ogilvies should have family jewels; but those are exactly the jewels that are almost always set in particular articles of ornament. The Ogilvies would seem to have kept theirs loose in their pockets, like coppers.

"Second item. Heaps and heaps of loose snuff, not kept in a horn, or even a pouch, but lying in heaps on the mantelpieces, on the sideboard, on the piano, anywhere. It looks as if the old gentleman would not take the trouble to look in a pocket or lift a lid.

"Third item. Here and there about the house curious little heaps of minute pieces of metal, some like steel springs and some in the form of microscopic wheels. As if they had gutted some mechanical toy.

"Fourth item. The wax candles, which have to be stuck in bottle necks because there is nothing else to stick them in."

Could anything possibly link these curious facts? Father Brown at once proposes a linking hypothesis.

"I think I see the connection," said the priest. "This Glengyle was mad against the French Revolution. He was an enthusiast for the *ancien régime*, and was trying to re-enact literally the family life of the last Bourbons. He had snuff because it was the eighteenth century luxury; wax candles, because they were the eighteenth century lighting; the mechanical bits of iron represent the locksmith hobby of Louis XVI; the diamonds are for the Diamond Necklace of Marie Antoinette."

Not that the priest thinks this at all likely, or other fantastic hypotheses that he spins. But now we start to get more evidence.

"Items five, six, seven, etc.," he [the policeman] said, "and certainly more varied than instructive. A curious collection, not of lead pencils, but of the lead out of lead pencils. A senseless stick of bamboo, with the top rather splintered. It might be the instrument of the crime. Only, there isn't any crime. The only other things are a few old missals and little Catholic pictures, which the Ogilvies kept, I suppose, from the Middle Ages – their family pride being stronger than their Puritanism. We only put them in the museum because they seem curiously cut about and defaced."

The priest gets perturbed, worrying that some desecration is afoot. He asked that the grave be opened and the body examined. When this is done, it is found that the corpse is headless, and suddenly through a chance remark about going to the dentist the priest has the answer. Israel Gow had been hired by the late earl, when the latter discovered that the former was a totally – obsessively – honest man. For a small errand, Gow had mistakenly been given a sovereign instead of a farthing, and instead of returning it or keeping it, he had returned with the change – nineteen shillings, eleven pence, and three farthings. The earl made Gow his heir, promising him all of the family gold, and Gow had been systematically removing the gold and nothing else. The earl's head was missing, for Gow was about to remove the gold from the fillings. And when this was done, the head was to be returned to the grave. Indeed, later that morning, the detective "saw that strange being, the just miser, digging at the desecrated grave, the plaid round his throat thrashing out in the mountain wind; the sober top hat on his head."

Now, let us take this apart. We have a number of items that need explaining, the snuff and so forth. A hypothesis is proposed, the love of the ancient regime. Given its truth, the items are now understandable. But it is clearly not the right answer. If we wanted, we could check on the late earl's interests, and surely find that he had no interest in the ancient regime. On the other hand, if we had found that he had a library of works on the topic, then we might take the hypothesis more seriously. But even then, we would not know why the snuff was out of its box (and the location of the box itself). One can likewise dismiss other hypotheses, which in any case would appear increasingly implausible as the new information to be explained came on board. Then the "just miser" hypothesis is raised. The priest has independent evidence for this – the story of the hiring of Israel Gow – and it leads to a successful prediction, namely, that the head is being plundered for its gold and once emptied will be returned to its grave. This hypothesis is highly plausible. It binds together many disparate pieces of information. It explains them exactly. It is consistent with what we know from elsewhere. It can be tested through the predictions that it makes.

The kind of argumentation that we are considering is generally known today as the "argument to the best explanation" (Lipton 1991). (Another name for the process comes from the American pragmatist Charles Sanders Peirce, who called it "abduction.") One has a series of hypotheses to compare, and by thinning them according to their success and plausibility on various grounds, like consistency with other material or theories, one brings the number down to the best remaining explanation. A somewhat stronger version is where you bring in material from various areas under the same hypothesis, thus simplifying the explanation significantly. And perhaps the strongest version is where you start to make all sorts of predictions of phenomena that you did not know already and these predictions start to pan out.

The last versions of the argument to the best explanation, those that stress unification, are sometimes known as "consiliences," this being the terminology of the nineteenth-century historian and philosopher of science William Whewell (1840). "The Consilience of Inductions takes place when an Induction, obtained from one class of facts, coincides with an induction, obtained from another class. This Consilience is a test of the truth of the Theory in which it occurs." The idea is that somehow, if a hypothesis is true – tells us about the real world – then various facts or other claims follow from it, and will keep doing so. And there is a kind of feedback process here. As the hypothesis leads to new information, so its derivations themselves confer a kind of probability upon the hypothesis. A false hypothesis would simply not keep working as well. At the least, you would have to keep adding to it, and thus lose the elegance or simplicity that scientists prize so much. Father Brown supposed that the miser was taking the gold; the hypothesis explained the strange phenomena, including the expected restoration of the head to the corpse; and these phenomena in turn made the priest's hypothesis highly plausible.

The Consilience of the *Origin*

Turning to Darwin himself, we know that he set out deliberately to construct a consilience, and felt with some reason that he had succeeded (Ruse 1999a). (Figure 2.4). This was part and parcel of his professionalism as a scientist, discussed in the last chapter. In the second half of the *Origin*, following the prescriptions of Whewell, whom he knew personally, Darwin tried to explain in a range of areas – instinct, paleontology, biogeographical distribution, systematics, anatomy, embryology, and more. Given his success at using an evolutionary hypothesis to explain so many disparate factors, he felt that he had succeeded in showing evolution to be true – to be a fact and not a theory, in the language of today's critics. In each area, we see him working toward an argument to the best explanation, and then rolling the whole system together in one consilience.

Take, for instance, the area of biogeographical distribution, always a major interest of Darwin, since it was so vital in triggering his own move to evolutionism. Darwin continued to be fascinated by the inhabitants of islands and discussed them in detail in the *Origin*. Basically, one has two hypotheses – that organisms were created (naturally or supernaturally) elsewhere and brought to the islands, or that they were created on the

Structure of Darwin's Theory

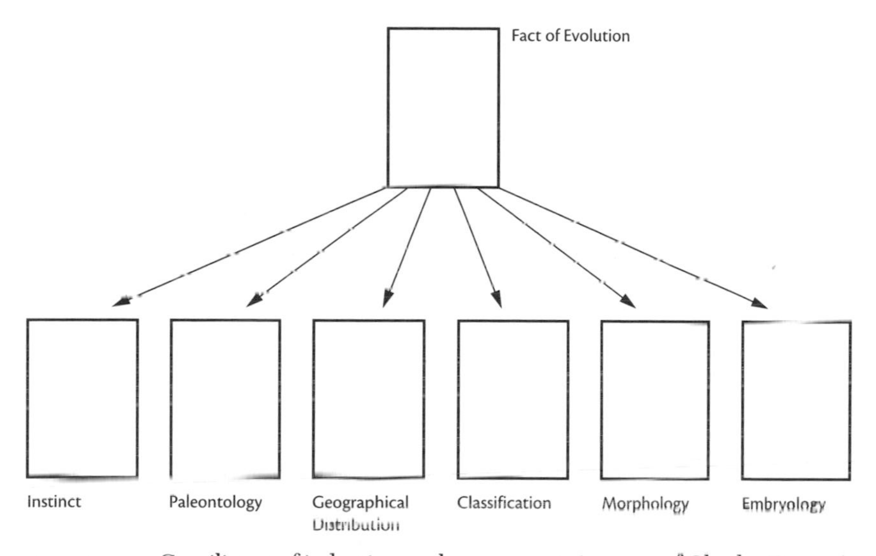

FIGURE 2.4. Consilience of inductions – the argument structure of Charles Darwin's *Origin of Species*.

spot. And, if on the spot, were they created naturally or supernaturally? One supposes that they could have been created elsewhere, but counting against this hypothesis is the fact that elsewhere in the world, more often than not, one does not find members of what would be the original group. Why does one not find those who did not make the trip? Why was it only the movers who survived? There are also major problems with supposing that organisms were made supernaturally on the spot. One would expect that a rational being like God would have put those organisms best suited in the same sorts of places. But nothing of the sort happens. Rather, we get very distinctive clusterings of organisms, starting with the fact that islands tend to have fewer varieties than mainlands, and that some organisms, like mammals – bad travelers over large expanses of sea – tend not to be found on islands. More than this, the inhabitants of islands resemble the organisms on their closest mainland, rather than on mainlands far away.

Numerous instances could be given of this fact. I will give only one, that of the Galapagos Archipelago, situated under the equator, between 500 and 600 miles from the shores of South America. Here almost every product of the land and water bears the unmistakable stamp of the American continent. There are twenty-six land birds, and twenty-five of these are ranked by Mr. Gould as distinct species, supposed to have been created here; yet the close affinity of most of these birds to American species in every character, in their habits, gestures, and tones of voice was manifest. So it is with the other animals, and with nearly all the plants, as shown by Dr. Hooker in his admirable memoir on the Flora of this archipelago. The naturalist, looking at the inhabitants of these volcanic islands in the Pacific, distant several hundred miles from the continent, yet feels that he is standing on American land. Why should this be so? why should the species which are supposed to have been created in the Galapagos Archipelago, and nowhere else, bear so plain a stamp of affinity to those created in America? There is nothing in the conditions of life, in the geological nature of the islands, in their height or climate, or in the proportions in which the several classes are associated together, which resembles closely the conditions of the South American coast: in fact there is a considerable dissimilarity in all these respects. On the other hand, there is a considerable degree of resemblance in the volcanic nature of the soil, in climate, height, and size of the islands, between the Galapagos and Cape Verde Archipelagoes: but what an entire and absolute difference in their inhabitants! The inhabitants of the Cape Verde Islands are related to those of Africa, like those of the Galapagos to America. (397–8)

As Darwin concluded, these sorts of facts "admit of no sort of explanation on the ordinary view of independent creation," but fall straight into place as soon as one supposes an evolutionary origin.

Move on from a best explanation in one area and look at the whole consilience (feeling free to go beyond Darwin to later discoveries). We have the largest range of biological phenomena that is explained by the fact of evolution, and that in turn confirms the fact itself. Start with the fossil record. Not only is the record roughly progressive, which is what we would expect were evolution true, but it also shows the kinds of sequences that we expect from evolution (Bowler 1976; Benton 1987, 2006). The most famous example is the horse. The modern horse was introduced into the Americas by Europeans, but the horse itself had a long evolutionary history in the New World. When he went to America in the 1870s to lecture, Huxley made the horse the centerpiece of his discourse, relying on magnificent finds from the West. He was able to trace, with specimens, the evolution of our single-toed quadruped back to a form that ran on four toes. He made a prediction: "in still older forms, the series of the digits will be more and more complete, until we come to the five-toed animals, in which, if the doctrine of evolution is well founded, the whole series must have taken its origin" (Huxley 1877, 90). Within two months, just such a five-toed animal (Hyracotherium, more popularly known as Eohippus) was found!

> Said the little Eohippus
> I think I'll be a horse
> And on my middle finger
> I'll run my daily course.

Huxley's presentation gave the impression that the evolution is a single-pathed affair, from beginning to end. Now we know that it is much more complex, with huge amounts of bushing and divergence (Simpson 1951). This is what we expect. We know also, particularly from the study of teeth, that there were various fluctuations, as lifestyles changed, from browsing to grazing and (for some) back again. Again, this is what we expect. Evolution is messy, as many other examples of change also show. These examples can be multiplied again and again, and moreover we have some wonderful cases of transitional organisms – the so-called missing links between the organisms of the past and those of today, or between one form and another, both in the past. Most famously, there are the links between reptiles and birds, especially one discovered (in full form) just after the *Origin* was published: Archaeopteryx (Feduccia 1996). I will talk about this in Chapter 4. And later I will talk about another link, occurring in the case of humans. Lucy, *Australopithecus afarensis*, lived more than three million years ago and was fully upright (although perhaps not as good a walker as we), yet she had an ape-sized brain (she did not have an ape brain, rather one of the size of an ape). Beyond the brute, if not yet human (Johanson and Edey 1981). The links or potential links multiply. There are many more similar cases. All of these are explained by the fact of evolution, and not properly by other hypotheses (if such there be). Conversely, they give credence to that which is doing the explaining, namely, the fact of evolution.

The same is true of other areas. Take, most strikingly, morphology. The most arresting fact is that organisms of quite different species have isomorphisms – those already-mentioned "homologies" – between the parts of their bodies, their limbs especially, even though the ends or functions of these parts are quite different. Most famous is the front limb of the vertebrate. Humans have arms and hands for lifting and grasping, horses legs for running, bats and birds wings for flying (although the ordering of bird and bat bones are quite different), moles paws for digging, seals fins for swimming, and so forth. Why would this all be so? What Darwin called "descent with modification" gives the answer.

> We have seen that the members of the same class, independently of their habits of life, resemble each other in the general plan of their organisation. This resemblance is often expressed by the term "unity of type;" or by saying that the several parts and organs in the different species of the class are homologous. The whole subject is included under the general name of Morphology. This is the most interesting department of natural history, and may be said to be its very soul. What can be more curious than that the hand of a man, formed for grasping, that of a mole for digging, the leg of the horse, the paddle of the porpoise, and the wing of the bat, should all be constructed on the same pattern, and should include the same bones, in the same relative positions? (434)

Another area picked up by Darwin was embryology – it is an area still made much of today. Why do we find that the embryos of organisms very different as adults – humans and chickens and dogs, for instance – have embryos that are very similar? Simply because they are descended from common ancestors. There is no other good reason.

> The points of structure, in which the embryos of widely different animals of the same class resemble each other, often have no direct relation to their conditions of existence. We cannot, for instance, suppose that in the embryos of the vertebrata the peculiar loop-like course of the arteries near the branchial slits are related to similar conditions, – in the young mammal which is nourished in the womb of its mother, in the egg of the bird which is hatched in a nest, and in the spawn of a frog under water. We have no more reason to believe in such a relation, than we have to believe that the same bones in the hand of a man, wing of a bat, and fin of a porpoise, are related to similar conditions of life. No one will suppose that the stripes on the whelp of a lion, or the spots on the young blackbird, are of any use to these animals, or are related to the conditions to which they are exposed. (439–40)

Vestigial organs are another piece of evidence for the evolutionary case. Many, if not all, organisms carry functionally useless features that seem to have functioning counterparts in other (more or less similar) organisms. Famous examples are the wings of flightless birds, the limb remnants of "limbless" vertebrates like snakes and whales, the eye sockets of eyeless fish, and in humans such things as the appendix and the coccyx, that bit of tail at the bottom of our backbones. Why do these exist? On any theory that makes adaptation totally ubiquitous, they would not have been created. But on a theory of evolution, they follow naturally as relicts and evidence of the past – of past ancestors, that is, shared with organisms that still (as did the ancestors) use these features for their own adaptive ends. "Rudimentary organs may be compared with the letters in a word, still retained in the spelling, but become useless in the pronunciation, but which serve as a clue in seeking for its derivation. On the view of descent with modification, we may conclude that the existence of organs in rudimentary, imperfect, and useless condition, or quite aborted, far from presenting a strange difficulty, as they assuredly do on the ordinary doctrine of creation, might even have been anticipated and can be accounted for by the laws of inheritance" (455–6).

But Is It Really True?

The point is made. We have many different areas of biology putting up phenomena that are explained by the fact of evolution, and for which other rival explanations – especially that organisms were created somehow entire, on the spot where they are now found – are simply inadequate. In each case, we have inferences to the best explanation; combined, they make a consilience that points strongly to the truth of the fact of evolution. More than this, the explanation fits well with the rest of science as we know it. In fact, in this respect, one surely has one of the marks of a truly great consilience, namely, that one is finding and explaining things unknown at the time of the consilience, or even things that were or would be considered difficult or contrary. Why, for instance, do we find fossils of the same Triassic reptiles on continents throughout the world, or southern beeches (Nothofagus) in both Chile and New Zealand, or conifers (Araucarias) both on the South American continent and in Australasia? Simply because these organisms were on land that once was whole, contiguous, rather than split as it is now into many different land masses separated by great seas and oceans (Cox and Moore 1993).

Not that these are the only cases of predictions of new-and-altogether-unsuspected phenomena. Until recently, no one would have looked for homologies between organisms as different as, say, fruitflies (*Drosophila*) and humans (*Homo sapiens*). Indeed, eminent evolutionists denied explicitly that one would find them. Now, however, we know that at the molecular level the similarities are mind-boggling (Carroll, Grenier, and

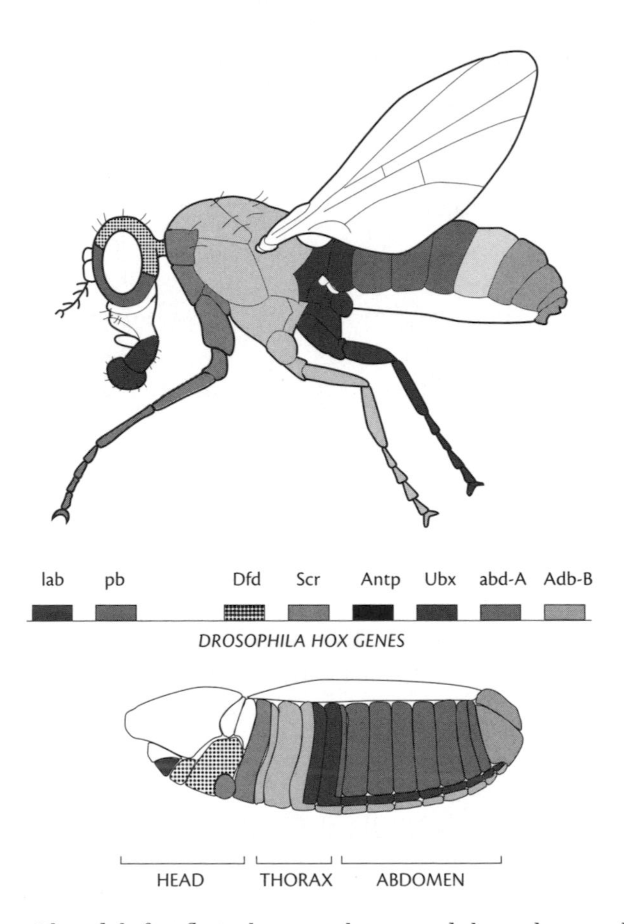

FIGURE 2.5. The adult fruitfly is shown at the top and the embryo at the bottom. The *Hox* genes, shown in the center, sequentially order the manufacture of the body parts. From Sean B. Carroll, Scott Grenier, and D. Weatherbee, "Homeotic genes and the evolution of arthropods and chordates," *Nature* 376 (2001): 479–85. Reprinted by permission.

Weatherbee 2001). Most remarkable of all are certain so-called "homeotic genes." Structural genes are those carrying the information needed to make actual bodily products. Developmental genes carry the information needed to process the production of bodily products by structural genes. The homeotic genes belong to the second category, regulating the identity and order of the parts of the body – they tell the body where to put an eye and where to put a leg, and a mal-mutation would perhaps move an eye to where a leg might normally appear, or vice versa. A subclass of such

Fly Dfd	PKRQRTAYTRHQILELEKEFHYNRYLTRRRRIEIAHTLVLSERQUKIWFQNRRMKWKKDN	KLPNTKNVR
AmphiHox4	TKRSRTAYTRQQVLELEKEFHFNRYLTRRRRIEIAHSLGLTERQIKIWFQNRRMKWKKDN	RLPNTKTRS
Mouse HoxB4	PKRSRTAYTRQQVLELEKEFHYNRYLTRRRRVEIAHALCLSERQIKIWFQNRRMKWKKDH	KLPNTKIRS
Human HoxB4	PKRSRTAYTRQQVLELEKEFHYNRYLTRRRRVEIAHALCLSERQIKIWFQNRRMKWKKDH	KLPNTKIRS
Chick HoxB4	PKRSRTAYTRQQVLELEKEFHYNRYLTRRRRVEIAHSLCLSERQIKIWFQNRRMKWKKDH	KLPNTKIRS
Frog HoxB4	AKRSRTAYTRQQVLELEKEFHYNRYLTRRRRVEIAHTLRLSERQIKIWFQNRRMKWKKDH	KLPNTKIKS
Fugu HoxB4	PKRSRTAYTRQQVLELEKEFHYNRYLTRRRRVEIAHTLCLSERQIKIWFQNRRMKWKKDH	KLPNTKVRS
Zebrafish HoxB4	AKRSRTAYTRQQVLELEKEFHYNRYLTRRRRVEIAHTLRLSERQIKIWFQNRRMKWKKDH	KLPNTKIKS

FIGURE 2.6. Comparison of the proteins produced by the *Hox* genes of fruitflies and of vertebrates. See how similar we are to the flies! From Sean B. Carroll, Scott Greiner, and D. Weatherbee, "Homeotic genes and the evolution of arthropods and chordates," *Nature* 376 (2001): 479–85. Reprinted by permission.

homeotic genes contains "*Hox* genes," a group of genes to be found in bilaterians (organisms the same on both sides). They order the appearance of various bodily parts, and (for once nature taking the easy and obvious way) they seem to work in the sequence in which they are found on the chromosomes. In *Drosophila*, the *Hox* genes start up at the head, work down through the thorax, and so on to the end of the abdomen. Within these genes, one finds lengths of DNA, of 180 base pairs, that are used to bind the genes to other DNA segments that are part of structural genes. In other words, these "homeoboxes" make a protein (of sixty amino acids) – the "homeodomain" – that is the key to the *Hox* genes actually functioning in regulating the structural genes. What was truly astounding was the discovery of a homology between the homeodomains of *Drosophila* (fruitflies) and other bilaterans, from frogs through fish and mice to humans (Figures 2.5, 2.6). The flies' parts and the humans' parts go back to the same processes. How else can one explain this other than through the hypothesis of evolution? There is no evidence that this is the only way in which the information for parts could be coded. Like isomorphic bones, the past tells the story.

Really a Fact?

Judged against the kinds of criteria and practices that we normally apply and use when making inferences, the evidence for the fact of evolution is very, very solid. But is it enough? Is it really enough to allow us to speak of evolution as a fact? Whewell was so opposed to evolution that supposedly he would not allow the *Origin* on the shelves of the library at Trinity College, Cambridge (where he was Master). He was being false to his philosophy here. He thought that a consilience was a guarantee of truth.

This was bound up with his beliefs (as an Anglican minister) that God would not let us be deceived about matters in cases where we had done exactly what we should to find the truth. But clearly, Whewell goes too far at this point. Obviously, it is not logically impossible that things happened another way, however strong the evidence. In the sexual assault case, it is not logically impossible that aliens somehow could replicate the DNA of any human, and do the dirty work. Just as it is not logically impossible that a bunch of Calvinists, seeking to discredit Father Brown and the Catholic Church, might have set up some elaborate deception to make us think that Israel Gow was the person responsible, because he was seeking all of the gold of the late earl.

But in none of these cases are the alternative scenarios very reasonable, and we can set about checking them out. In the assault case, we want some evidence that aliens are around and that this is the sort of thing that they are habitually doing. My information, gleaned mainly from reading the *National Enquirer* at supermarket checkouts, is that aliens are chiefly in the business of kidnapping Elvis. It could be true that the DNA was deposited by an identical twin, but one can check the existence of such a relative. There are other (somewhat disgusting) possibilities for the existence of the DNA – some were suggested by the defense in a rape-murder trial in Toronto a few years back – but again it is possible to check and rule these out. Rule them out until, in the opinion of a jury or any other thoughtful person, the assault is proven "beyond reasonable doubt" – as happened in the Toronto trial.

Similarly in the evolution case. I suppose that logically it is possible that this world of ours is an artificial laboratory set up by Andromedans, who are checking on questions of deception and the functioning of artificial organisms and so forth. It is logically possible that we humans were created and are programmed to think as we do, even though in fact we are going through motions in a world created (let us say) six thousand years ago. Finally, it is logically possible that junior Andromedans are getting Ph.D.s for studying how we humans function in various situations. ("Let us manipulate them into a world war and see how they behave.") But there is absolutely no reason to think that any of this is so, and much to think that at best it is a philosophical (or Hollywood) fancy. For a start, we know how difficult it is to create any kind of artificial life, let alone a whole world teeming with it. And how are the Andromedans checking on us and communicating their results back home? And so forth.

So the answer is that if you insist that evolution be logically, absolutely proven so that there is no possibility of any kind of fantasy being the alternative, then you cannot have it. But the second part of the same answer is that never in real life do normally people demand this kind of proof – we can never have it, and we never feel the need of it. So if you are satisfied that John Fitzgerald Kennedy really was a human being, a male, who became president and was assassinated in 1963 – that he was not a transvestite for whom a substitute was killed in Dallas, and that the real Kennedy doesn't live in Texas or Moscow (your choice) with his/her husband or wife (your choice again) – then you really ought to accept evolution as a fact.

Methodological Naturalism

But is this not to make one assumption that is in its way fatal to the whole enterprise? The obvious (as well as historical) alternative to evolution is some kind of religious solution. Whether or not it be strictly Christian, the alternative is that God somehow, at some time, intervened miraculously – non-naturally or supernaturally – and created organisms as we now find them. This alternative has simply been ruled out by fiat. Just before Darwin, the naturalist Philip Gosse (1857) (a member of that odd sect known as the Plymouth Brethren) suggested that everything was created by God – fossils and all – at one fell swoop to make the world as it is. Have we not simply dismissed this without argument? Today we have enthusiasts for so-called Intelligent Design, who suppose that the nature of the living world is so complex that blind, unguided law simply could not do the job, and so there has to have been some kind of intervention by the Designer, Whoever he (or she or it) might be (Behe 1996; Dembski 1998a, b). Are they not being dismissed without argument?

Well, actually it is not quite true that we have ignored entirely the supernatural alternative. Certainly under normal understandings (that is, Judaeo-Christian understandings) of the Creator, we have been eliminating that option. (Not the option that there is such a Creator, but that the Creator did it all at once, miraculously.) Vestigial organs really are rather silly – deceitful even – if God did it all at once, and if that God is all-powerful and so forth. It is true that we have not tackled the specific arguments of all of those who opt for miracles – those who claim that a natural (lawbound) solution will never explain complexity. This topic will be left until later, when we are more ready to look at it. But what about the simple claim that we ought to take seriously the bare possibility of some kind of non-natural (miraculous) intervention that made things as they are? The idea that we ought to take seriously something like the Andromedans, but out of the regular course of nature? Of course, this requires some kind of leap of faith, but have we not already conceded that some kind of leap of faith is needed to get to the fact of evolution, even in the natural world? You cannot rule out, logically, the possibility of an alternative scenario.

An immediate response here is that science is precisely that against which one defines faith – what faith requires is what science refuses to give, so by definition evolution cannot be a matter of faith. But this is too quick a response. The real issue is why it is reasonable to go with the scientific position, and to reject the miracle-invoking position, however you label the two. Why should we – even the religious person – think that blind law is the way to explain this world? Why should the scientific person be committed to what we might call "methodological naturalism" – the working assumption that all physical events can and must be explained by laws – even if, as a person, one is not committed to ontological or (what we might call) "metaphysical naturalism" – the philosophical or religious belief that there is something more than material nature. In other words, why insist on explaining the world through blind law when you might well believe that something else might stand behind everything?

The answer given by the scientist, including the evolutionist, appeals to the pragmatic. Methodological naturalism works! As Thomas Kuhn (1962) says about paradigms, because scientists have persisted in taking a methodologically naturalistic approach, problems that hitherto seemed insoluble have eventually given way to solutions. Take an example from biology. For many years – indeed, ever since Darwin – there was much debate about how insect sociality could have evolved. How is it that the worker ants, for instance, devote their whole lives to the nest, despite the fact that they do not reproduce themselves? People had no answer but did not give up. They persisted, and finally in the early 1960s, William Hamilton (1964a, b), a graduate student at the time, provided an explanation, showing how one can explain insect sociality in terms of individual genetic selfishness. (Briefly, the answer is that in the hymenoptera – the bees, the ants, and the wasps – females are more closely related to sisters than they are to daughters. Thus, they improve their genetic success by raising fertile sisters, rather than by raising fertile daughters. Later, I will spell out the details of this explanation.) The methodological naturalist says that this is a moral for us all: although there are indeed many unsolved problems, notably the origin of life, past experience suggests that these problems will be solved eventually by a methodologically naturalistic approach. Therefore, one should persist, no matter how improbable the finding of a solution seems today.

Is this the final word? Come back to where we came in, to the thinking of Alvin Plantinga. He thinks that for the Christian this is simply the wrong way to go about things. A Christian is a theist, believing that God is no remote being, but involved intimately with His creation, all of the time. Methodological naturalism rules out this whole perspective.

> First and most important, according to serious theism, God is constantly, immediately, intimately, and directly active in his creation: he constantly upholds it in existence and providentially governs it. He is immediately and directly active in everything from the Big Bang to the sparrow's fall. Literally nothing happens without his upholding hand. Second, natural laws are not in any way independent of God, and are perhaps best thought of as regularities in the ways in which he treats the stuff he has made, or perhaps as counterfactuals of divine freedom.

Plantinga at once adds that because God always stands behind His laws, all of the time, "there is nothing in the least untoward in the thought that on some occasions God might do something in a way different from his usual way – e.g., raise someone from the dead or change water into wine" (Plantinga 1997, 149). In an important sense, law and miracle are of the same logical type, and hence the switch from one to the other is to be expected rather than otherwise. Of course, the critic is right if he or she complains that now Plantinga is committed to what we might call "science stoppers." This is a consequence of abandoning the ubiquity of methodological naturalism. But as Christians, have we any right to assume the nonexistence or impossibility of science stoppers?

> The claim that God has directly created life, for example, may be a science stopper; it does not follow that God *did not* directly create life.... Clearly we cannot sensibly insist in advance that whatever we are confronted with is to be explained in terms of something *else* God did; he must have done *some* things directly. It would be worth knowing, if possible, which things he *did* do directly; to know this

would be an important part of a serious and profound knowledge of the universe. The fact that such claims are science stoppers means that as a general rule they will not be helpful; it does not mean that they are never true, and it does not mean that they can never be part of a proper scientific theory. . . . It is a giant and unwarranted step from the recognition that claims of direct divine activity are science stoppers to the insistence that science must pretend that the created universe is just there, refusing to recognize that it is indeed *created*. (Plantinga 1997, 152–3)

At this point, a scientific argument will not suffice. We must fight theology with theology. And this has been done by the philosopher of science and Catholic priest Ernan McMullin. Casting the discussion specifically in terms of our thinking about the origin of new organisms, with respect to the Genesis story McMullin writes as follows:

The issue, be it noted, is not whether God *could* have intervened in the natural order; it is presumably within the power of the Being who holds the universe at every moment in existence to shape that existence freely. The issue is rather, whether it is antecedently *likely* that God would do so, and more specifically whether such intervention would have taken the form of special creation of ancestral living kinds. Attaching a degree of *likelihood* to this requires a reason; despite the avowed intention not to call on Genesis, there might appear to be some sort of residual linkage here. In the absence of the Genesis narrative, would it appear likely that the God of the salvation story would also act in a special way to bring the ancestral living kinds into existence? It hardly seems to be the case. (McMullin 1991, in Hull and Ruse 1998, 712)

McMullin's point is that Plantinga is only arguing that miracles are, at some level, as likely as laws because he has in the first place made a fairly literalistic reading of the Bible. But this now raises a second objection: it is by no means obvious to one working from a Christian position that one must agree that God works almost indifferently through law and through miracle. First, it is only if one has already made a priori a fairly literalistic reading of the Bible that one would think that God's miracles are going to be as frequent as Plantinga suggests. If one interprets, let us say, the story of Noah not so much as a literal case of marine history but as an allegory about human nature – not the least important part coming after the Flood, when God has cleansed the world and Noah celebrates by getting stone drunk and Ham misbehaves – then the whole question of frequent miracles by God becomes rather more problematic. Second, one can (as McMullin and others point out) make a distinction between the order of nature and the order of grace. That is, between what is known as "cosmic history" and what is known as "salvation history." To quote McMullin again:

> The train of events linking Abraham to Christ is not to be considered an analogue for God's relationship to creation generally. The Incarnation and what led up to it were unique in their manifestation of God's creative power and a loving concern for the created universe. To overcome the consequences of human freedom, a different sort of action on God's part was required, a transformative action culminating in the promise of resurrection of the children of God, something that (despite the immortality claims of the Greek philosophers) lies altogether outside the bounds of nature.
>
> The story of salvation is a story about men and women, about the burden and the promise of being human. It is about free beings who sinned and who therefore *needed* God's intervention. Dealing with the human predicament 'naturally', so to speak, would not have been sufficient on God's part. But no such argument can be used with regard to the origins of the first living cells or of plants and animals. The biblical account of God's dealings with humankind provides no warrant whatever for supposing that God would have brought the ancestors of the various kinds of plants and animals to be outside the ordinary order of nature. (725–6)

One has no expectations because of God's use of miracle in a certain special set of events that God will be using miracles as frequently or indifferently as Plantinga suggests.

All in all, therefore, one can say that Plantinga has not made his case about the likelihood of miracles occurring as often as laws for the Christian theist. And with this, we can bring to an end the discussion of this chapter. Evolution is at the center of a powerful consilience. By any normal understanding of the terms, evolution is a well-established fact. It is logically possible that evolution is not true, but it is not reasonable to believe this. And questioning this on theological grounds does less than justice to the science, and most probably to religion also.

CHAPTER THREE

The Origin of Life

An impenetrable barrier to science and a residue to all attempts to reduce biology to chemistry and physics.

K. R. Popper (1974)

This was the judgment of the important Austrian-British philosopher Karl Popper about the origin of life. He believed that it set a problem that has not been answered and that, for all we know, may never be answered. If this is from the pen of a man who loved science and respected its power and scope, you can imagine the comments from the other end of the spectrum. Of the claim that "life arose by naturalistic means," Alvin Plantinga (1998) writes: "This seems to me for the most part mere arrogant bluster; given our present state of knowledge, I believe it is vastly less probable, on our present evidence, than is its denial" (p. 685). This too is the position of the atheist-returned-to-the-fold Anthony Flew. He finds "improbable" the possibility of a natural origin of life. "I have been persuaded that it is simply out of the question that the first living matter evolved out of dead matter and then developed into an extraordinarily complicated creature" (Wavell and Iredale 2004).

If things are this bad, then perhaps our best strategy might be to avoid the problem altogether. As we turn now from discussing the fact of evolution to discussing the path of evolution, why not plunge right at the point when life is up and running? Why not leave origins to others? There is precedent for doing this. Charles Darwin's discussion of the problem in the *Origin of Species* always puts me in mind of Sherlock Holmes's response in the story "Silver Blaze." Upon being asked if there were any points of note, he replied: "The dog that barked in the night." "But the dog didn't bark in the night." "Precisely!" "But Darwin didn't discuss the origin of life in the *Origin of Species*." "Precisely!" He knew that he had no answer and that getting into a discussion of the topic would lead only to tears, so he stayed away from it altogether.

However, that is cheating a little – certainly it would be today. If we want to give a proper account of the history of life – the path of life – we surely must look at the topic of life's origin. That will be the topic of this chapter. Then, in the next, we can turn to the issues that Darwin himself covered. Although I am reserving full discussion of Darwin's mechanism, natural selection, to a later chapter, in this and the next chapter I shall introduce discussion of pertinent causal issues as needed to make sense of the story. One word of warning. As we set out to look at the origin of life, at once we encounter a paradox. No self-respecting evolution textbook today avoids the topic, but the truth is that evolutionists are really not at all qualified to talk on the subject – one needs masses of biochemistry and like knowledge really to get involved, far more than that possessed by the average student of life's history. So to a great extent we have to rely on the work of others. First, though, we should ask about our inquiry: What is it of which we seek to find the origins? What is this life that we want to explain?

Vitalism and Its Critics

Most obviously, the answer to this question is that life is some sort of substance. Rocks never had it. A dead cow had it and has now lost it. You, my reader, have it, and I have it. We have life; the cow had it and then it went; the rock was ever lifeless. But what kind of substance is this life? As an undergraduate, my wife majored in animal and poultry science. In one never-to-be-forgotten course, a living pig was brought into the classroom, weighed, taken out and shot, and then the carcass returned and cut up, and the parts weighed and examined. The students dutifully noted down the size of the various bodily parts and compared them to estimates that they had made when the porker was still alive and squealing; but never were they expected to note down the weight of the living substance and to compare it to the weight of the dead substance, seeking (to use a term often employed in such discussions) the "life force." Such forces, if they exist, have no weight.

Not that this is necessarily a bar to thinking in terms of a substance. Most of us, in everyday life, follow philosophers like Plato and Descartes in thinking that the mind in some way is a force, and a substance – to each of us, it is as real as our hands and arms (and a lot more real than, say, our pancreas, that most of us could not locate or tell its function, if we were challenged). Yet no one thinks that the mind has weight. You do not lose ten pounds when you fall asleep. Perhaps we might say that the life force is a bit like a mind – not so much a conscious mind, for most of us do not think that trees have thoughts, but something akin to a mind that is working always and subconsciously. A kind of animating (or vegetating) force.

This was the position of the great Greek philosopher Aristotle (1984), who was also a serious (and, as we now realize, very good) marine biologist. For living organisms, he thought there was some kind of soul – not in the Christian sense (he wrote four hundred years before Jesus) or in any theological sense at all, but in the sense of something that makes for life and action. It is not just a question of the material out of which organisms are formed, for "we are inquiring not out of what the parts of an animal are made, but by what agency. Either it is something external which makes them, or else something existing in the seminal fluid and the semen; and this must either be soul or a part of soul, or something containing soul." He thought that it was something that exists "in the embryo itself" – a kind of life force, making for vitality, for the very fact of living and breathing and being. We may not see it directly, but just like the mind itself, it is there and functioning.

This idea proved extremely influential, and it persisted in one way or another right down to the twentieth century. Indeed, it was around the beginning of this last century that – probably driven by dissatisfaction with the materialistic tendencies of modern biology – a number of people began to champion the view that a return to Aristotle is needed. Life itself – especially evolving life itself – demands more than a purely naturalistic approach. These people wanted to escape from a purely materialistic viewpoint – one that saw nature as driven by blind mechanisms – and they reverted to supposing that it is life forces that animate and push life forward. In Germany, leading this new "vitalism" was the embryologist Hans

Driesch. He argued that we need to invoke something that he termed an "entelechy." In France, the leading figure was the philosopher Henri Bergson (1907) – part Jewish but later to grow close to Catholicism. (He wanted to convert, but, under the Nazi occupation, bravely refused to renounce his Jewish heritage and died from a chill caught queuing for identity papers.)

Bergson was a deeply committed evolutionist. There was no compromise on gut positions. Nevertheless, Bergson judged that all then-existing theories of evolution – all then-existing purely mechanistic theories of evolution – were inadequate. For Bergson, as for so many others from Charles Darwin to Richard Dawkins, it was the complexity issue that was crucial. He thought that a mechanistic view destroys the holistic view that we need of the organism. "The real whole might well be, we conceive, an indivisible continuity. The systems we cut out within it would, properly speaking, not then be *parts* at all; they would be *partial views* of the whole. And, with these partial views put end to end, you will not make even a beginning of the reconstruction of the whole. . . . " This decompositional approach spells the end of trying to understand the essence of life. "Analysis will undoubtedly resolve the process of organic creation into an ever-growing number of physicochemical phenomena, and chemists and physicists will have to do, of course, with nothing but these. But it does not follow that chemistry and physics will ever give us the key to life" (Bergson 1911, 32–3).

We need something that will give direction to evolution. We need something that will make for complexity and the like. Bergson located this creative power – something he christened the *élan vital* – in the life force or impetus possessed by all living things. "This impetus, sustained right along the lines of evolution among which it gets divided, is the fundamental cause of variations, at least of those that are regularly passed on, that accumulate and create new species" (pp. 92–3). The impetus, the *élan vital*, is like consciousness, deciding on the best path to take and then trying to walk along it. There is direction and the influence of the end, but not so much from the end itself as from the consciousness of the end. Defining life as "a tendency to act on inert matter," Bergson continued: "The direction of this action is not predetermined; hence the unforeseeable variety of forms which life, in evolving, sows along its path. But this action always presents, to some extent, the character of contingency; it implies at least a rudiment of choice. Now a choice involves the anticipatory idea of several possible actions. Possibilities of action must therefore be marked out for the living being before the action itself" (pp. 101–2). Apparently, sight is just such a possibility of action, and that is why complex eyes have evolved several different times.

Bergson's ideas were highly influential. But for all this, we must admit how greatly out of tune with modern science was such thinking. The problem was not so much that the *élan vital* was unseen or directly unknowable. Twentieth-century science in particular is loaded with the unseen and directly unknowable. What are we to make of electrons, with their complementary qualities of particles and waves? What ultimately separated the *élan* from the electron was that whereas the latter notion is very useful – indispensable – the former notion is not. It gives the impression of explanatory power, but it is not embedded in laws and cannot be used for prediction or unification or any of the other epistemic demands that one makes of the unseen entities of science. One can do just as much without the *élan* as one can do with it. The palaeontogist G. G. Simpson (1949) put his finger on things: "Granting, as any reasonable person must, that there is an important difference between life and non-life, you may, if you wish, call the different behaviour of matter in life 'vitalistic,' but this accomplished nothing and means nothing that was not already obvious. It is an example of the naming fallacy to call this an explanation or a contribution to evolutionary theory" (p. 125).

This was so before the days of the DNA molecule and the double helix, and is even more so after. The *élan vital* was of no help to biologists in cracking the genetic code. Molecular biologists have found no place for the *élan vital* as they struggle to follow the development of the organism from the macromolecules of nucleic acid to the finished adult. And worse even than this, the *élan* seemed to commit one to some ideas deeply antithetical to modern science. There is full agreement with Francis Bacon that science should eschew forces that impose direction on the course of events – "teleological" forces. Yet, for all that Bergson claimed that he did not want to give consciousness to all living matter, that kind of teleology was precisely what he was introducing into science. And with this, he was attributing consciousness to organisms where such attribution just does not seem justified. If the trilobite, say, or the plant, did not have consciousness in some wise, then how could it choose to go in one direction rather than another?

Organization

If life is not a substance, what then is it? To use a hackneyed term, a paradigm shift was in the wind. Perhaps life is not really a substance at all. Although people agreed with Bergson when he seized on the issue of complexity as the issue of real significance, many thought that this might point the way to the solution rather than merely to the problem to be tackled. As the secrets of the cell and of heredity were being uncovered, the tremendous organization involved was what struck many. It is not so much that we have different materials – more and more, people were finding that the materials are basic chemical compounds – but the way they are put together, the way – and here again, of course, we go back to other aspects of Aristotle's thinking – things are put together and function. The difference between a clock that works and a clock that does not is not a question of "clock force," but a matter of one being put together properly while the other is not. The one is organized properly, and the other is not. Many interested in the nature of life, therefore, were drawn to dropping substance talk and taking up the language of organization.

One highly influential figure here was the great evolutionist of the early twentieth century J. B. S. Haldane (1949). He argued explicitly that life is basically a question of organization. Take a play by Shakespeare. It is as much a verbal experience as something visual. "Be, food, if, love, music, of, on, play, the," is a list. "If music be the food of love, play on!" is the start of great literature. Of course, the content matters. "It is important to know this, as it is important to know that life consists of chemical processes. But the arrangement of the words is even more important than the words themselves. And in the same way, life is a pattern of chemical processes." Haldane stressed that these are processes of a particular kind – in particular, processes that are able to keep themselves going and to replicate in some wise. "This pattern has special properties. It begets a similar pattern, as a flame does, but it regulates itself in a way that a flame does not except to a slight extent. And, of course, it has many other peculiarities. So when we have said that life is a pattern of chemical processes, we have said something true and important." What is true and important is that life is not a simple substance – a thing. It is rather a matter of the way in which elements are put together.

Of course, we cannot be talking about just any kind of organization. One might think of a crystal as organized, but it is not living. It has to be organization of a special kind, namely, the kind that we associate with living things. And what is it that we associate with living things? Clearly, reproduction and (if we are evolutionists) the ability to change and diversify. So the organization has to be put in terms of being something capable of self-generating and of maintaining itself – even if we agree that a crystal can generate itself, it cannot maintain itself. Nor can it evolve. Of course, this is probably going to give us some borderline cases. Thunderstorms might seem to get close to qualifying. Anyone who lives in Florida and has been in the path of a tornado will not think this a joke. Closer to biology, is a virus something living or not? It perpetuates itself only parasitically on other organisms. Probably the right response is to say that borderline cases are what we get in the real world – especially in the real world of evolution. Tornados, thank goodness, do not live and persist forever. So they are disqualified. Viruses are a judgment call, and therefore perhaps make the case rather than destroy it.

None of this is to trivialize the significance of life or to say that now its origin is at once explained – anything but. Rather, it is to take the unnecessary mystery out of the concept of life and recognize it for what it is rather than for what it is not or might be. The dead pig weighs no less than the live pig, but its organization has been disturbed – its brain and its nerves and so on have been messed up by the slaughterer's bullet, and so it no longer works. It is no longer alive. Life is a matter of functioning and organization rather than substance. And with this shift in emphasis, we shall now see that the origin-of-life question became transformed – at once, more complex and yet more promising.

Spontaneous Generation

Why is this? Start with the fact that if we think of life as a matter of organization rather than as a substance, then we have a tool that helps us to explain the history of origin-of-life studies as well as a light on today's discussions. From the time of Aristotle and before, people were speculating on whether all life was eternal or whether it appeared at some particular time, and if the latter, whether all at once or sequentially and repeatedly (Farley 1977; Fry 2000), and if sequentially and repeatedly, whether miraculously or naturally, that is, according to unguided laws. The Jews had it all laid out as a series of miracles, sequentially, over a short period of time. The ancient Greeks, however, tried out various natural or quasi-natural options. The already-mentioned Empedocles (fifth century B.C.) thought that it was all a matter of the recombination of the four elements (earth, fire, air, and water) under the forces of strife and love, making random bodily parts. Then these parts come together into functioning organisms – by chance. Aristotle, however, perhaps not entirely consistently with the rest of his philosophy, inclined to the natural appearance of life in one step – the idea we encountered in Lamarck, so-called spontaneous generation. He thought the absolute beginning of life something analogous to the beginning of life through reproduction. Clearly important was the belief that now you do have (or do not have) the life force, and now you do not have (or do have) the life force. In an important way, the life force and spontaneous generation are different sides of the same coin. No one then thought that organization comes about spontaneously. However, life forces do lend themselves to spontaneous appearance. Things are in one state or another, because of this substance – this thing – and so it is plausible to think of life beginning at one specific instant. It is as though the light bulb, which was dark and nonfunctioning, were suddenly turned on.

With its mirror image of life force, the idea of spontaneous generation was to have a long shelf life. For all that it is not a part of the Jewish creation story, some version of it was endorsed by many Christian philosophers. This was in large part because of Augustine, who believed that God was outside time, and that hence that to Him the thought of creation, the act of creation, and the product of creation are as one. God created seeds of potentiality that became activated and developed through time (McMullin 1985). Although coming from a very different perspective, a millennium and a half later, this kind of thinking appealed to the early evolutionists. They were more deist than theist, and so their motivation was primarily that of seeing God's intentions worked out through unbroken law. But like earlier thinkers, they thought that now you do not have life, and now you do. It appears in a flash. Lamarck supposed that primitive lifelike worms come out of ponds filled with warm mud, the action of lightning and so forth sparking the beginnings. So did Erasmus Darwin.

> Then, whilst the sea at their coeval birth, Surge over surge, involved the shoreless earth; Nursed by warm sun-beams in primeval caves Organic life began beneath the waves.... Hence without parent by spontaneous birth, Rise the first specks of animated earth. (Darwin 1803)

But even as they appeared on the scene, the evolutionists were starting to go against the tide. In the seventeenth century, the physician Francesco Redi blew the cover (literally and metaphorically) on many putative instances of spontaneous generation. He showed that they were the results of insects laying eggs, which later hatch. Meat that is covered with muslin or some such cloth, thus preventing the settling of flies, remains maggot-free. "Although it be a matter of daily observation that infinite numbers of worms are produced in dead bodies and decayed plants, I feel, I say, inclined to believe that these worms are all generated by insemination" (Farley 1977, 14). A century later, the Italian Lazzaro Spallanzani boiled broths in sealed flasks. He showed that once life has been destroyed, it never reappears. Not that any of this was ever definitive – many argued, for example, that Spallanzani had proved nothing because his experiments destroyed everything that might have made for life in the first place.

The real blow came, as everyone knows, in the middle of the nineteenth century, when the French chemist Louis Pasteur pointed to the end of these kinds of arguments. In a series of experiments detailed in textbook after textbook, he boiled sugared yeast to kill off the live contents. Then he showed that the treated material remained sterile unless and until it was recontaminated. Pertinently, he was able to demonstrate that these are not one-off or fluke results. They hold again and again in different conditions. Even exposing the material to the outside air makes no difference. If the methods of exposure are such that they prevent contaminants from entering (openings that are long and thin and curved), there is no new appearance of life. It is only when nonsterile substances (including freely circulating air) are permitted to infect the inner material that fermentation begins again. Life demands life.

The empirical evidence was important but not definitive. Interestingly, Pasteur himself always inclined to think that spontaneous generation from nonorganic materials is possible. And others, including some prominent evolutionists who thought that spontaneous generation was obligated by their transmutationary thinking, continued to favor the idea. But its days were virtually finished. The vitalists may have resuscitated thinking about life forces, but this only served to heighten the fact that such thinking is profoundly out of step with the ways of modern science. As the empirical evidence was telling against spontaneous generation, so the philosophical critique was telling against life forces. Organization rather than substance is the key to life, and in a post-Darwinian world, organization means putting things together bit by bit, functioning all of the way, rather than having one, instant, all-inclusive step.

The Oparin–Haldane Hypothesis

Darwin himself sensed this, speculating (in a letter written ten years after the *Origin*) that life's beginnings were a gradual process. But it is two people particularly, in the 1920s, who are credited with the ideas that started things moving in the direction of gradual development of life from nonlife: first, the Russian biochemist Alexandr Oparin (1924), and then (not much surprise here) the already introduced J. B. S. Haldane (1929). There were differences in their thinking. Haldane, for instance, thought that because viruses are somewhere between the nonliving and the living, they might be the key to origins. But, essentially, both men postulated the same developmental picture of the beginning of life. You start with simple molecules and you build up, bit by bit, until you have fully functioning life. Let us go step by step through, not so much their own reasonings, but the stages they postulated for life's origins, seeing where we stand on these issues today (Orgel 1994; Brack 1999; Knoll 2003).

The first move must be to make the ultimate building blocks of life – starting with nonorganic molecules, you must make things like amino acids, the components of vital macromolecules that both make and drive the components of the cell. This means that you need the right conditions. Suppose that the original conditions were much as they are today, with a 20-percent-oxygen atmosphere. Haldane (who also worked as a chemist) pointed out that the needed molecules simply could not have formed and persisted. With so much oxygen around, they would have broken down at once. In other words, the original atmosphere must have been very different from what we have today. Specifically, one had to have what chemists call a "reducing" atmosphere – that means one without free oxygen but with free hydrogen.

As it happens, back when Haldane was writing, folks thought that this was no problem. If anything, it was taken as an encouraging sign that life might have started naturally here on Earth. Go back to the beginning stages of planet Earth. Our home is about four and a half billion years old. Initially, it was molten. It started to cool, slowly, and over the next billion or so years, the oceans formed. The atmosphere supposedly was oxygen-free.

Instead, one had such gases as methane (CH_4), ammonia (NH_3), carbon dioxide (CO_2), and hydrogen sulphide (H_2S). Just as pertinently, it would have been an atmosphere that did not block ultraviolet radiation. For this barrier, you need an ozone (O_3) layer, and such did not exist. All in all, it seems to have been just what was needed for the formation of the right molecules – amino acids and so forth. Moreover, if they did appear, they would not at once be degraded out of existence.

But could they form? This is the second part of the Oparin–Haldane hypothesis. Among the amino acids, there are some twenty key varieties that go into making the proteins, the components of cells and the enzymes that make cellular processes work. (Enzymes themselves are not part of processes, but they are needed for such processes, aiding them, for instance, by picking up two different molecules and joining them.) As is well known, shortly after the middle of the twentieth century, it seemed that the solution to the origin-of-life problem was falling quickly into people's hands, thanks to a famous experiment by a graduate student named Stanley Miller (1953), working in Chicago in the lab of Nobel Prize–winner Harold Urey. In a container full of a reducing atmosphere, Miller mixed the appropriate quantities of inorganic compounds (including methane and ammonia), and he then set an electric spark to discharge intermittently through the container and its contents. This latter was to imitate the lightning supposed to be a regular feature of the early Earth. And lo and behold, a week later Miller found that he had a handful of the right kinds of amino acids. Nor was this a chance result and finding, because since then others have continued in the same vein and have produced even more of the needed organic molecules, and not just the parts of proteins, but also some of the required building blocks for the nucleic acids, those macromolecules that carry the genetic information needed to build organisms and to pass on life from one generation to the next.

Unfortunately, those early celebrations proved premature. For a start, since Miller did his experiments considerable doubt has been thrown on the assumption that the atmosphere of the early Earth was in fact reducing (Bowring and Housh 1995). Many now think that it contained considerable free oxygen, or at least enough oxygen in forms that would prevent the kinds of reactions that Miller was able to duplicate. Perhaps the necessary molecules could not have formed after all – at least, not as originally supposed. As it happens, serious though this objection undoubtedly is, there is a growing number of scientists who doubt that it is a definitive refutation of the possibility of appropriate molecule synthesis. On the one hand, there is considerable evidence that the needed building blocks do in fact get formed elsewhere in the universe. Meteorites often contain just the required materials. Moreover, there is strong evidence that the bombardment of the Earth by meteorites in its early days was a significant factor in the formation of the planet. So it is possible that, although the required molecules are now synthesized down here, the earliest molecules were imports.

On the other hand, many suspect that the site of the formation of the earliest life forms may not have been the open sea – what Darwin (speculating on life's early history) called a "warm little pond" and what Haldane called an "organic soup" – but rather the deep-sea vents up through which bubbles magma (molten rock), thus forming the plates on which the continents are moved around the globe. The required basic molecules are down there; there is no free oxygen; and there is lots of energy in the form of heat from the Earth's center. No need of the sun and its light, or lightning, or whatever, to drive reactions. This may all seem a little implausible, but supporters of this suggestion argue that it is more surprising than incorrect (Wächtershäuser 1992). Their faith is bolstered by the fact that some of the most primitive and earliest-known organisms (to be mentioned in Chapter 4) are themselves denizens of hot-springs and deep-sea vents. This does not prove anything in itself, but it is at least suggestive.

RNA World?

A more serious worry is about the stage after the building-block formation. Here we run into a classic chicken-and-egg situation. Suppose you do have amino acids. How are these going to be linked up into long chains, thus making proteins? And worse, even if you do get proteins, how are you to get a functioning cell (or precursor)? For the making of proteins, you need nucleic acid to program. Conversely, if you focus on the nucleic acid, then how do you get the apparatus to make proteins – the constituents of the cell and the enzymes that drive the processes of the cell? (Figure 3.1)

There are some hints about how nature might have linked up building blocks into long chains. Actually, the linking itself is no big issue. This sort of thing happens quite regularly. More problematic is getting molecules to stay linked. Chains tend to degrade, and you are right back

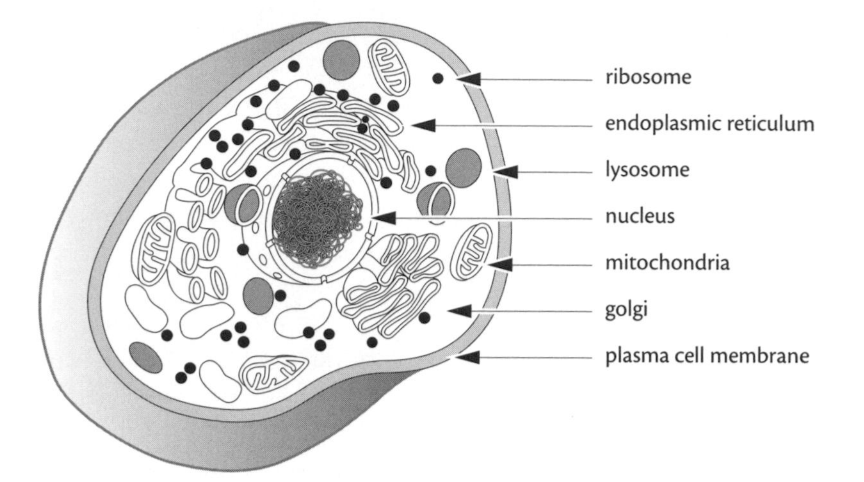

FIGURE 3.1. Cross section of an animal cell. ribosome: RNA used to make proteins; endoplasmic reticulum: transport system; lysosome: digestive enzyme, used for recycling parts; nucleus: cell center containing chromosomes; mitochondria: energy sources; golgi: shipping department taking material out of cell; plasma cell membrane: outer shell.

to where you were when you started. Here, however, nature does offer some clues. Naturally occurring clays may be significant causal factors. Organic molecules adhere to clays, and can build up chains, while at the same time resisting the urge to break apart. Already experimenters have shown how quite long chains can be formed in such (presumably analogous to natural) conditions. This is an answer to only part of the problem, of course. You need more than just linking. You need linking in some order that leads to functioning, and you need the right macromolecules to keep things going – Haldane's move beyond mere organization to replicating organization.

What of the need to complement proteins with nucleic acids? A protein – many proteins – can do nothing in the long run without nucleic acids in place. Conversely, nucleic acids can apparently do nothing without the appropriate proteins. Some suspect that what we have here is a situation akin to dry-stone bridge building. If you tried to put up a bridge without cement, then as you moved the sides closer to the middle, at some point everything would topple over and down. What you must do is build a support from wood or earth, on which you lay the stones. When finished, you can remove the support, and the stones pushing against each other will stay in place and won't fall. Could it be that at the beginning of life there were supports, now long removed? One suggestion has been that crystals were the key factor – they can replicate almost exactly, with just a few blemishes or errors. Perhaps organic molecules piggybacked on crystals and eventually were able to go it alone and self-reproduce (Cairnes-Smith 1986).

Others prefer to explore avenues opened by some major discoveries of about twenty years ago (Gilbert 1986; Pace and Marsh 1986; Szostak, Bartel, and Luisi 2001). It turns out that ribonucleic acid (RNA) – the acid that in most cells is used to copy the information from deoxyribonucleic acid (DNA) and then to make proteins – can itself form enzymes that can in turn drive reactions. It is known that ribonucleic acid can itself act as the ultimate carrier of genetic information – it does so today, in some simple organisms. Could it not therefore have been that at one early point we had an exclusively RNA world, with these molecules making the enzymes to self-replicate? Now we need to raise the question of order. If our early RNA molecules were fabricated higgledy-piggledy, then how could they function to self-replicate? Some classic experiments in the early days of molecular biology showed that in the right circumstances, naked RNA molecules (given the apparatus to reproduce) can mutate and compete against each other, thus setting up a selective situation, which in turn leads to evolution – evolution in an adaptive direction. (The molecules move toward more rapid reproduction, even at the cost of dropping a lot of their length, if this length proves superfluous). Today, therefore, we have people trying to bridge the gap. Somehow we have got to get the RNA molecules to build themselves naturally, under their own steam, competing against each other to improve. Once this is done, it should be possible to get RNA not only to build itself, but also to copy itself and (given mutation) to evolve in the direction of ever-improved adaptive functioning.

An Impenetrable Barrier?

Obviously, even if this all works out nicely, we hardly yet have life as we know it. Very significant steps remain. What about the shell or coat of the cell? You need some kind of naturally occurring globe into which everything can be put and sheltered from outside forces. Oparin spent much of his life working on these problems, as did the American researcher Sidney Fox (1988). They worked (with some success) to find how organic molecules can be made to form self-contained spheres (like the outer shells of cells). At the same time, they strove to show how these spheres could maintain themselves and do the sorts of things that cells need to do – bud off to form other spheres; keep and even promote differences between the inside and the outside; and select some compounds to cross over from the outside to the inside (while barring others). Of course, these workers found that not every compound is of equal worth, and they were able to show that some of these compounds can be precisely the kinds of molecules (like ribonucleic acid molecules) that one would expect to have been preserved and cherished in new or proto-cells.

Even if we can get things wrapped up in their own containers, there is still much more to be done. But, for now, let us leave the story. With life in existence, we can defer until the next chapter further empirical discussion of the path of evolution. Return now to the skeptics at the beginning of the chapter. Let us not dodge the problem or try to weasel out of a challenge. If one is a Darwinian, then one is committed (qua Darwinian) to a natural origin of life, and one agrees that this is going to be part of the overall picture – part of the Darwinian's overall picture, not someone else's. So we have to ask ourselves again about the problem – the origin of life – and the extent to which today we have the solution. We have seen that there has been a shift in the philosophical underpinnings of this question. Virtually no one today – certainly no one active in the scientific community – thinks that this is a matter of substances. Life is not dead material plus something else. Life is better thought of as a matter of organization and functioning. We have also seen the science – the work done on the formation of the vital macromolecules and on their functioning and so forth. Now we ask whether the opinions of the philosophers – Popper and Plantinga and others – are well taken.

Let us defer until the final chapter (on religion) those like Flew who have opted out of naturalism, and consider in their own right the arguments against today's naturalistic thinking. One problem is that neither Popper nor Plantinga spell out in any detail their worries and objections. In Popper's case, one clue to his discomfort comes from his talk of the origin-of-life problem as a barrier to reduction. What could this mean? Unfortunately, "reduction" is one of those extendable terms, a bit like "God," that means different things to different people (Nagel 1961; Ayala 1974). Leave aside the chemical sense used earlier – an environment without the corrosive influence of oxygen. Turn to the philosopher's usage.

Sometimes reduction means the possibility of deducing one area of science from another – "theoretical reduction." The relationship between molecular genetics (the DNA molecule) and Mendelian genetics (the Mendelian gene) is generally taken to be a case of theoretical reduction, for the latter is deducible from the former. There is no conflict between the two fields, it is just that one is more powerful than and absorbs the other. In the origin-of-life case, there is no theoretical reduction. Clearly no one is deducing biology from physics and chemistry here, but the origin-of-life problem is not really setting out to do this. One is trying to see how life appeared according to (and only according to) the laws of physics and chemistry, but there is no intention of providing a formal theory – certainly not now, and probably never. Sometimes reduction means explaining in terms of the ever-smaller – genes good, molecules better. Certainly this kind of reduction – generally known as "methodological reduction" – is at work here. But it is hard to see this as a problem or a fault. Because of methodological reductionistic practices, no one today thinks that spontaneous generation is plausible, and a good thing too.

Sometimes reduction means eliminating one substance in terms of another – "ontological reduction" – as one might try to explain consciousness only in material terms. Popper disliked this (Popper and Eccles 1977). He was a "dualist," thinking that thought is something essentially different from material entities. (Presumably Bergson was likewise at least a dualist. I referred to his "holistic" perspective, which is a generic term for all those who deny ontological reduction.) However, regardless of your attitude toward consciousness as such, with the shift from thinking of life as implying vital forces to seeing it as implying organization, the ontological reduction issue in this context is put to one side. Inasmuch as one has order in biology, it is as much physico-chemical order as it is biological order. What is the order on a DNA molecule – physico-chemical or biological? The question really does not make sense. It is the order itself that is important, and all scientists concur on this.

Popper seems to have given no knock-down argument showing that the origin-of-life problem is insoluble. What about the other end of the spectrum? What of the disdain and contempt of someone like Plantinga? All one can say is that if a scientist behaved this way toward Plantinga's subject, philosophy, he would be rightly incensed. The so-called ontological argument claims to derive God's existence from His perfection. It was first formulated by Saint Anselm in the eleventh century and then taken up by Descartes in the seventeenth. One is invited to consider the nature of God, and seeing (in Descartes' version) that He is perfect, one concludes that He must exist, since it is more perfect to exist than not to exist. Try this analogy. Of the claim that "the ontological argument for God's existence is more subtle and significant than is realized by the average undergraduate," the biologist Joe Bloggs writes: "This seems to me for the most part mere arrogant bluster; given our present state of knowledge, I believe it is vastly less probable, on our present evidence, than is its denial." That puts things in reasonable perspective. I myself do not believe that the ontological argument is valid. I know that it is a lot more subtle and significant than is realized by the average undergraduate. If one is not prepared to make any effort to understand what is happening in the laboratories, then one has no warrant to make comments or judgments, certainly not negative ones.

Plantinga's prejudices are no more significant than are Popper's. But let us not end our discussion prematurely. We have made some real effort to look at the science, so let us put aside our natural irritation at the philosophers' cavalier dismissals. Given the science – and for the sake of argument, grant the science – where do we truly stand today? Should the Darwinian be worried or perturbed or start to question his or her basic position? On the one hand, we can certainly say that the origin-of-life problem has not been solved (Orgel 1994). We can certainly say that it is hugely more difficult than anyone thought in the years before molecular biology. Early evolutionists had no conception of the issues. On the other hand, we can certainly say that a lot of important and pertinent work has been done and that it is ongoing. Are we close to a solution? It is hard to answer that question until we have the hindsight that stems from a solution, but I suspect that everyone's educated guess is that we have still a long way to go – certainly a long way to go if we want the full picture.

So is this a cause for despair and a call to start searching, if not for miracles, then at least for something radically different? Do we need yet another paradigm shift and new laws? This is roughly the position of the popular writer on science/religion issues Paul Davies (1999). He denies that life's beginning was supernatural, but he does seem to want to leave open the possibility that there was a little bit of something more. "I do believe that we live in a bio-friendly universe of a stunningly ingenious character" (p. 20). He is looking for "new philosophical principles" that will have "immense philosophical ramifications," and he seems to think that information will be important. But without being unduly rude, one would like to know what the cash value of this call really is. As we have seen, there has been a philosophical shift from a search for substance to a search for order, and if that does not include information – the message of the double helix – one does not know what would. If these "new philosophical principles" are not a call for vitalism in new and friendlier clothing, one would like to have some idea of what they are – or if not what they are, what they are going to do in a way that is not being done already.

Enough of talking about others. What should we conclude? I want to say two things. First of all, let me return to the strategy of naturalism – methodological naturalism, that is. I want to remind you of how powerful and reasonable a strategy that is. If you are firmly committed to the belief that natural explanations can be found for all physical (including organic) phenomena, then you are going to think that natural origins of life is a reasonable idea, no matter what the gaps. The answer is not yet at hand, certainly not fully at hand, but this is the nature of the beast. You keep going until you succeed. But, in a case like this, is naturalism justified? Is one not simply arguing in a circular fashion, hauling in the conclusion that one wants? Do not discount the powers of indoctrination and prejudice. Today it would take a brave scientist indeed to admit that he or she was going to invoke miracles. I can just imagine the comments on the next grant application. Naturalism may be a good strategy generally, but perhaps sometimes Plantinga is right. The time comes to contemplate "science stoppers" – events beyond the normal course of nature.

What can one do but repeat the arguments given before. Speak first at the general level. To be a methodological naturalist is not unreasonable or a "leap of faith," in the sense that we normally understand faith. To believe that Jesus died on the Cross for our sins requires faith. I am not saying that this is a stupid belief – anything but; however, I would say that it is not a belief that can be justified by reason or the evidence. It is beyond reason in some sense. Indeed, there are religious thinkers – Søren Kierkegaard, notably – who think that faith in some sense has to be absurd. And there is truth in this. It is absurd to think that the death of a man two thousand years ago makes possible my eternal salvation. Absurd, but not stupid. Naturalism is a different fish. It is something that is based on evidence and the success of the past. You go beyond the evidence, but not into the absurd in the sense just described. You have had success in the past by "toughing it out" with difficult problems, by refusing to invoke science stoppers. And that means it is reasonable to tough it out in the future. When the people in London ridiculed their early attempts to build a DNA model, Watson and Crick did not give up, say "science stopper" and slip off to the pub across the road. They worked and worked until they had the molecule right – the double helix – and then they slipped off to the pub across the road (Watson 1968). Methodological naturalism works – and works and works again. So do not give up. Try and try again.

This turns us to the level of the specific. What about the particular case of the origin of life? Is this not a special case that makes the naturalistic approach highly unpromising? Not at all! Here the very brief history given earlier can help us to put things into perspective. Had we had the DNA model back in the time of Aristotle, and were we no further along some two and a half millennia later, then perhaps some doubts – some collywobbles – would be understandable, if not excusable. But it is only in the last hundred years or so that the magnitude of the problem has really started to become apparent, only in the last seventy years or so that the right or promising philosophical approach has started to dominate, and only in the past fifty years or so that the biochemical processes and information have opened up and made possible any powerful attack on the problem. And since then time has not stood still. We have just had a list – a litany – of moves and successes. The answers are not yet all in. One would not expect this. The problems are massive. Better modesty now than apologies later. The history of the all-too-slick spontaneous generation idea warns us of this. One has a progress report – a lengthy progress report – but no final document and conclusion. The time has not yet come for congratulations. But neither has the time come to commit intellectual hari-kari.

Does the progress report truly inspire confidence? One can certainly say that advances have been made, and that one has no sense that the researchers yet feel that they have come to a stop or a barrier. There is much ongoing work, for instance, about the ways in which one can synthesize nucleic acids in a natural or quasi-natural manner – getting the parts to link up and to stabilize and to start functioning. There is work about the possible conditions under which life might have started. Are those deep-sea vents really the clue to the process, or are they just hot air – or hot water? One really could not ask for much more at this time. Or if one does ask for more, in the manner of Popper or Plantinga (or Davis), one should know why one is asking for more and be prepared to specify what would satisfy. You cannot simply up the barrier because you do not want the solution to be found.

For this reason, the Darwinian need not fear that he or she is doing science that presupposes answers to an insoluble problem. The researchers' conviction that answers will come – that naturalistic answers will come – is not misplaced. The origin-of-life problem is not a threat. It is not an area of gloom and self-delusion. It is rather inspiring and exciting. There are Nobel prizes to be won. The critics and naysayers are wrong.

CHAPTER FOUR

The Path of Evolution

Evolution: The Fossils Say No!

Duane T. Gish is a Young-Earth Creationist. He believes that the Earth and its contents were created miraculously by God within the last ten thousand years or so, as described by Genesis. A Berkley-trained biochemist, he has been a very effective propagator of his position, debating with evolutionists on many college campuses, generally reducing his opponents to incoherent fury and delighting his audiences with his jokes and his ability to retort to every counter with a yet-more-outrageous and sweeping claim. He is the author of a little book – *Evolution: The Fossils Say No!* ("More than 150,000 copies sold!"). Gish knows all of the tricks. The very title of the book is a paradigmatic case of what philosophers call the fallacy of complex question. "Have you stopped beating your wife?" If you say "yes," then the response is "Why did you start in the first place?" If you say "no," then the response is "Why don't you stop?" Either way, you are caught, because the very question presupposes that you have been beating your wife. Likewise with Gish's book. The very title presupposes that the key evidence for evolution is the fossil record.

We know already that this is not true. For Darwin, the evidence of biogeography – those birds and reptiles in the Galapagos – was as (if not more) important, and morphology and embryology and the rest came in just behind. Even if there were no fossils at all, the case for the fact of evolution would be overwhelming. Gish is tricking us into arguing on his terms, and assuming that the fossils are the key factor. Once we have swallowed this, he is off and running. Does this then mean that the fossils are unimportant? Of course not! They help to make the case for the fact of evolution. But their real power is in tracing the path of evolution: phylogeny. Because of the fossils, we can speak about the course that change took down through the ages, from the beginning of this Earth of ours. Although, even here, let us jump in at once with a major qualification or supplement. The fossil record is not the only way in which we can ferret out the story of the past. There are two other avenues of discovery. One is through comparisons of the physical characteristics of organisms. The other is through looking at the molecules of organisms and working out the rates at which they change. More on these later, but first let us look at the fossils.

The Fossil Record

It was around the time of the Scientific Revolution in the sixteenth and seventeenth centuries that people first started to recognize that fossils are not just curiously shaped pieces of stone, but the remains of long-dead organisms (Rudwick 1972). This conclusion did not at once destroy people's faith in the creation stories of the Bible. No one knew exactly how old the fossils were, and in any case one could even use them to support the old beliefs. Fossil fish found in the Alps seemed clear evidence for Noah's Flood. Increasingly, however, these sorts of ploys failed to satisfy, and by the time that evolutionists appeared on the scene, the suggestion was being raised that the fossils tell us of a past of gradual change, from forms very different to today's organisms, including us humans. Paradoxically, although Lamarck was probably turned to evolution by fossils – he could not find the equivalent animals among the living and so assumed that they must have evolved to new forms – he did not think in terms of a roughly progressive fossil record supporting evolution. It was primarily the nonevolutionists, starting with Cuvier (1813), who began to work out the record, showing that there are definitely forms in the past that have no living equivalents, that as one goes back in time (as represented by levels deeper in the geological record) one sees forms ever-stranger, and that these older forms often combine the features of later organisms that are widely different. In some rough sense this was a progressive picture – fish and reptiles down among the lower levels, mammals up toward the upper levels, and humans right at the top.

For Cuvier, this was all somewhat coincidental, for he had a theory about organisms found in Europe having migrated from elsewhere; but by the late 1820s, there were an increasing number of people who were prepared to tie this in with changing climatic conditions, especially as the Earth cooled and got ready for humans. In 1828, referring to the Carboniferous, Adolphe de Brongniart wrote:

> During this first period, the atmosphere was freed of a part of its excess of carbon, by the vegetables which grew upon the land, which assimilated it, and were afterwards buried in the state of coal in the bowels of the earth. It is after this epoch, during our second and third periods, that the immense variety of monstrous reptiles began to appear, animals which, by their mode of respiration, are yet capable of living in a much less pure air than that which the warm-blooded animals require, and which, in fact, have preceded them at the earth's surface. (Bowler 1977, 24)

And so on and so forth, as the vegetation on Earth continued to remove carbon and finally made the globe a place that could be inhabited by mammals and then by humans.

As it happens, although Cuvier tried to explain organic form entirely in terms of function, German enthusiasts for some form of developmental picture (known as *Naturphilosophen*) seized on the record with joy, arguing that it shows the working out of the *Bauplan* or (what Richard Owen called) the archetype (Richards 1992) (Figure 4.1). For both functionalists and formalists, however, life had moved beyond a simple upward progression. Branching was crucial. Cuvier thought that there were four basic divisions – *embranchments* – within which variation can be seen; the formalists likewise thought that there are such divisions, which then reflect the variations on the different *Baupläne*. No one thought that one could go back to one basic plan, although some were starting to suggest that perhaps the divisions are not as unbridgeable as someone like Cuvier claimed.

For the nonevolutionists, each new form demanded a special creation, although by midcentury it was getting less and less easy to invoke the Deity. This just did not seem the right approach to a scientific problem. Darwin, inheriting this picture, saw problems for the evolutionist. There were gaps between forms at different levels of strata, but given the unlikelihood of fossilization, these perhaps were to be expected rather than otherwise. More troublesome was the fact that the record seemed to start suddenly

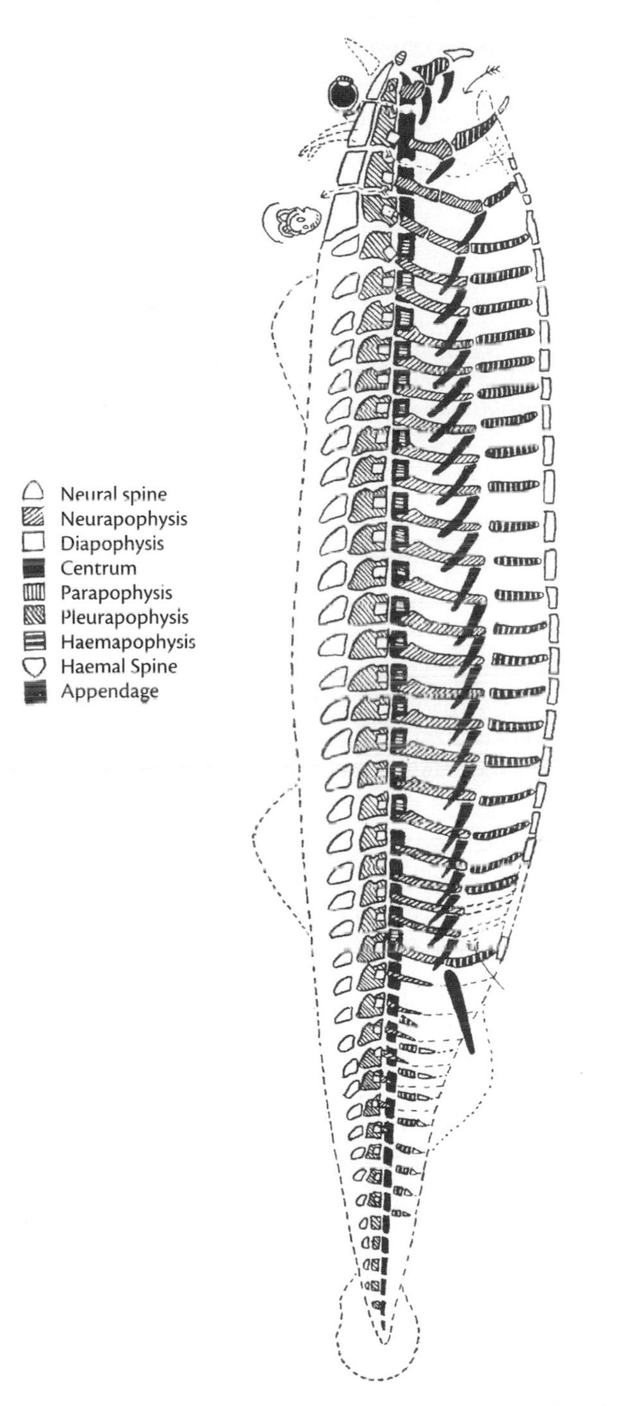

FIGURE 4.1. Richard Owen's picture of the vertebrate archetype (from his *On the Nature of Limbs*, 1849).

back in the Cambrian with full-fledged organisms having skeletons. If evolution were true, then one would expect to find earlier, more primitive forms right back to the barest traces of life. Darwin felt obliged to come up with an ingenious argument suggesting that the earliest forms of life had flourished where now there is sea and hence much pressure. Consequently, such forms would not be represented in today's record. More comforting for the evolutionist was the more or less progressive nature of the record after the Cambrian, with more primitive and basic organisms back in the past – basic organisms joining different extant forms. Other features were likewise significant to the evolutionist: just before the appearance of a new class, one often finds specimens that seem to foretell the newcomers. The nonevolutionist Louis Agassiz labeled these "prophetic types." "It seems to me even that the fishes which preceded the appearance of reptiles in the plan of creation were higher in certain characters than those which succeeded them; and it is a strange fact that these ancient fishes have something analogous with reptiles, which had not then made their appearance" (Agassiz 1885, 1, 393).

Really significant for Darwin and his followers was the branching nature of the record. For him, unlike Lamarck, branching was of the essence of his theory. Remember the Galapagos. Birds and reptiles had come to the islands as single groups and then, once there, they had divided and formed new species. The tree of life metaphor was as perfect for his theory as it was for the fossil record as then known. "As buds give rise by growth to fresh buds, and these, if vigorous, branch out and overtop on all sides many a feebler branch, so by generation I believe it has been with the great Tree of Life, which fills with its dead and broken branches the crust of the earth, and covers the surface with its ever branching and beautiful ramifications" (Darwin 1859, 130). Yet, although he discovered some important fossils while on the *Beagle* voyage, Darwin was not a paleontologist and generally left these matters to others. The *Origin* is not a work about paths. To get a good sense of what was known around that time, we must turn to others, one of whom – Richard Owen, an anatomist and rival of the Darwinians – gives a conveniently detailed picture of life's history (with the branches compressed into a single column) (Figure 4.2).

It can be seen that fish came first, then amphibians and reptiles, next birds and early mammals, and finally the forms that we know today. Similar pictures were drawn of the history of plants. Fast forwarding now to the present, the question becomes that of how much more we know

TABLE OF STRATA AND ORDER OF APPEARANCE OF ANIMAL LIFE UPON THE EARTH.

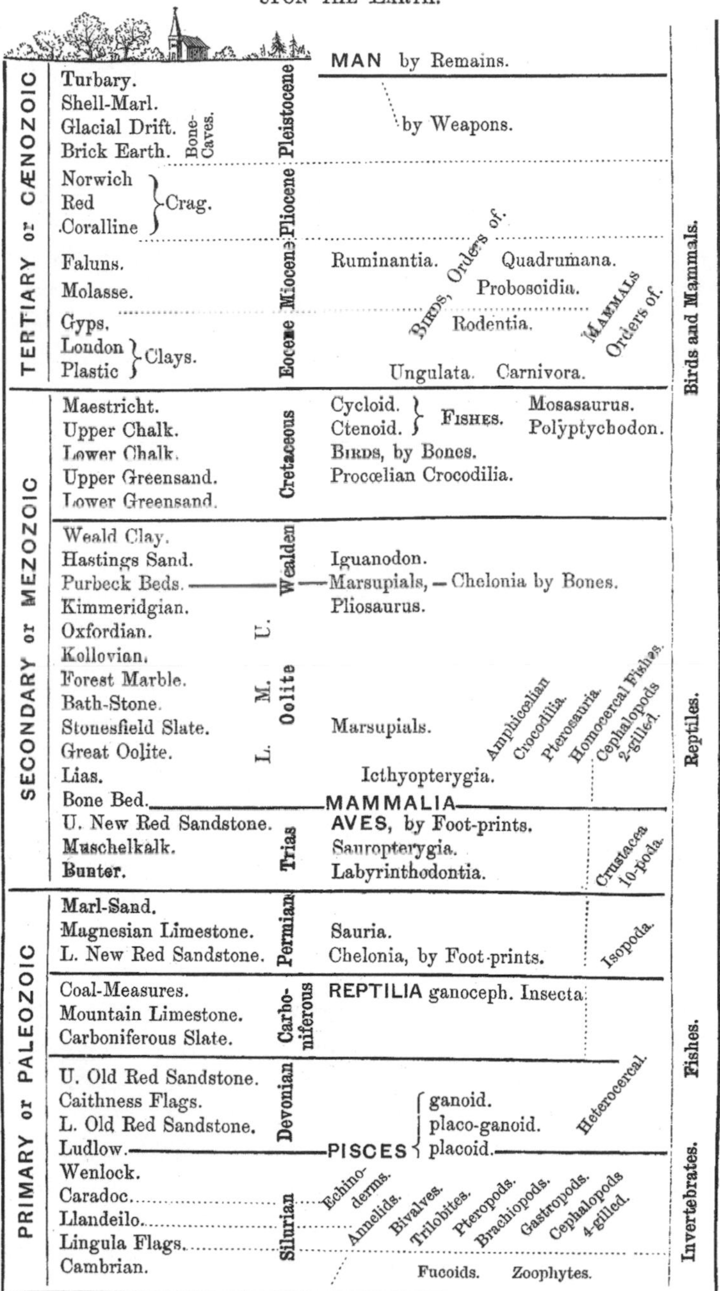

FIGURE 4.2. Richard Owen's picture of life's history (from his *Paleontology*, second edition, 1861).

today about life's history, and the extent to which our conception has had to be modified or changed. Part of the problem in answering this question is that, as common sense shows, there is something ambiguous about paleontology – you can never be sure that you have the last or first specimen of a group. Something new can always come along. People had thought that the fish called the coelacanth came from a group extinct for at least seventy million years – then, in 1938, just off Madagascar, a living one was hauled up from the sea. Likewise, if less dramatically, fossils have been unearthed that moved back significantly some of the lower limits marked by someone like Owen. Amphibians, for instance, have been pushed back from the early Carboniferous to the late Devonian, and mammals from the mid Jurassic to the late Triassic. So at one level, any answers about the completeness of the record have to be hedged. Nevertheless, this said, time has not stood still since the days of Darwin.

The Pre-Cambrian

One really major advance is that, thanks to methods of radioactive dating (whereby the rates of decay of certain elements are used as the basis for calculation), we can now assign fairly reliable dates to aspects of the record (Knoll 2003; Dawkins 2003a). We can indeed go all the way back to the origin of the Earth (about four and a half billion years ago) and the origin of the universe (about fifteen billion years ago) (Figure 4.3). Then, linking this up with the origin of life, really dramatic is the extent to which we now have evidence of life before the Cambrian. Indeed, it is thought that there are traces of life – hardly full-blown fossils – from about three and a half billion years ago. That means that, given that the earliest state of the Earth was undoubtedly very hot and molten, not something that any life could endure, life started here on Earth about as soon as it possibly could. We no longer need to get into Darwin's face-saving exercises, excusing the nonrecord of early life. Indeed, although very much is still unknown, we can already say a great deal about that very long period of life's history about which Darwin and his fellows were totally ignorant.

Start with one of the most important divisions between life forms – many would say *the* most important division, although we shall have more to say on this issue shortly – that between relatively simple organisms like bacteria (for instance, *E coli*, the bacterium found in our gut) and complex

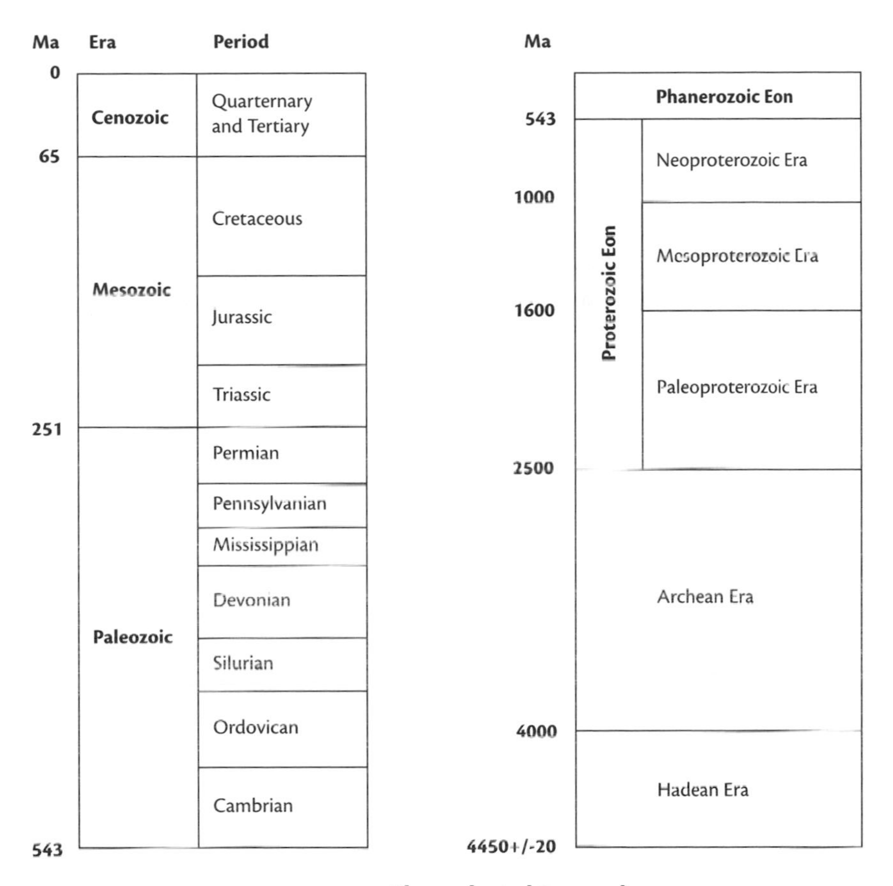

FIGURE 4.3. The geological time scale.

organisms such as animals and plants. Prokaryotes, examples of the former, are single-celled and have a fairly unstructured internal content, with nucleic acid loose in the center and with few other parts. Eukaryotes, the latter form, can be single-celled or a combination of many cells (like mammals and flowering plants, angiosperms) and have structured interiors – nucleic acids on chromosomes in a nucleus, and other interior cell parts like mitochondria and chloroplasts (used for energy metabolism, the former in plants and animals, the latter in plants only and used in photosynthesis). In 1967, the American researcher Lynn Margulis proposed the brilliant hypothesis that perhaps the eukaryotic cells were formed by incorporating prokaryotic cells – just as bees and flowers are bound in a symbiotic relationship, dependent on one another, so also are bigger cells which brought in smaller cells in such a mutually advantageous relationship (Margulis 1970, 1993). The mitochondria and the chloroplasts were once independent entities, but then got absorbed into other entities – absorbed but not destroyed, and able to go on functioning, now for the good of the whole rather than for the individual (refer back to Figure 3.1).

Although this was a highly controversial hypothesis when first proposed, today, thanks to the great advances in molecular understanding, the idea is widely accepted. Everyone knows that biological technology now lets you transfer genes from one organism to other, very different ones – from animals to plants, for instance – and there is some considerable evidence that something like this occurs in nature (Syvanen and Kado 2002). Microorganisms can snip out bits of DNA from one organism and then carry them over to other organisms, where they are reinserted, this time into a very different genome. In the case of putative eukaryotic cell evolution, the really convincing evidence was the finding that components like mitochondria and ribosomes have genetic sequences that are very similar to those of some prokaryotes, which of course is what one would expect if they themselves started life as prokaryotes. More than this, chloroplasts have DNA sequences similar to those of prokaryotes known as cyanobacteria, blue-green algae, that are themselves able to perform photosynthesis. (Tying things in even more tightly, cyanobacteria have metabolisms that are midway between fermentation, like regular prokaryotes, and respiration, like regular eukaryotes.)

What dates are we looking at? The earliest date that we have for the cyanobacteria is about 2.7 billion years ago (there is some fairly convincing direct fossil evidence and some completely convincing chemical materials that are produced by and only by cyanobacteria.) Around the same time, 2.7 billion years ago, there are also rocks that are rich in organic compounds, including molecules called "sterols" – their production is usually associated with eukaryotes (Brocks et al. 1999). (Cholesterol is today the sterol of most interest to us.) However, the first unambiguous fossil evidence of eukaryotes is much younger, about 1.2 billion years old. It is also around this time that we start to get evidence of eukaryotes being multicellular. There are a number of reasons that can be given to explain this gap between the two dates – that the earliest sterol-producing lifeforms were not full-blooded eukaryotes; that eukaryotes existed but were not laid down in the fossil record; that there was some breakthrough that suddenly gave the eukaryotes a major boost and made them move to a position of prominence. Probably the truth is a combination of these and perhaps other reasons. The popular supposition is that the eukaryotes got a real edge because of a significant evolutionary development, namely, sex. It is not strictly true to say that prokaryotes have nothing analogous to sex – *E. coli* join up and exchange pieces of DNA. But it is with the eukaryotes that sex as we understand it became possible and widespread, and regardless of why it was maintained in organisms – a topic to be discussed in Chapter 6 – the fact is that with sexuality and the rapid spread of useful genes in groups, the way was opened for massive and rapid innovation.

One significant factor in all of this was undoubtedly the rise in oxygen levels. Eukaryotes are oxygen users – they respire, that is, they get their energy from burning glucose in the presence of oxygen. Prokaryotes, on the other hand, generally are fermentation machines – they simply break down glucose and get their energy that way. As it happens, the two processes are not entirely different. You get the ability to respire by adding on steps to fermentation. This obviously is more evidence that the eukaryotes evolved from prokaryotes. But how was the move at all possible if early Earth history shows that there was no free oxygen – if, as we have already seen, early Earth history demands that there be no free oxygen? The most simple and obvious solution to this dilemma is that oxygen was produced just as it is today – by photosynthesis. In this process, thanks to the energy provided by the sun, oxygen is freed from carbon dioxide.

Do we have evidence that oxygen levels did start to rise about two billion years ago – around the time when one has the cyanobacteria, so that one might expect the evolution of the more strongly oxygen-producing eukaryotes? We do indeed. The clue here comes from the geological record, especially what happens to iron. Simply, iron without oxygen does not rust; iron with oxygen does rust.

> Some of the most compelling evidence for oxygen scarcity on the early Earth comes from gravel and sand deposited by ancient rivers as they meandered across Archean and earliest Proterozoic coastal plains. Pyrite [FeS_2 – fool's gold] is common in organic-rich sediments, forming below the surface where H_2S produced by sulfate-reducing bacteria reacts with iron dissolved in oxygen-depleted groundwaters. . . .
>
> The same is true of two other oxygen-sensitive minerals: siderite (iron carbonate, or $FeCO_3$) and uraninite (uranium dioxide, or UO_2). Neither of these minerals is found today among the eroded grains that make up sediments on

coastal floodplains, but both occur with pyrite grains in river deposits older than about 2.2 billion years. (Knoll 2003, 96)

Conversely, we find that after 2.2 billion years, we get the deposition of minerals that can form only in the presence of iron. The bright red sandstones of the Grand Canyon are a stunning example. "These rocks – called red beds, in the button-down parlance of geologists – derive their color from tiny flecks of iron oxide that coat sand grains. The iron oxides form within surface sands, but only when the groundwaters that wash them contain oxygen. Red beds are common only in sedimentary successions deposited after about 2.2 billion years ago" (Knoll 2003, 97). Before this date, there cannot have been more oxygen than about 1 percent of today's levels; after this date, there was at least 15 percent of today's levels. The way was being prepared for the rise of the eukaryotes.

The Cambrian and After

Notwithstanding the arrival of the eukaryotes, the record does not give evidence of an immediate or steady rise in life's complexity. Today, the start of the Cambrian is usually put at about 545 million years ago, rather later than was thought for many years. It was not until one or two hundred million years before this that life really started to pick up pace, and even here, although there is widespread evidence of more sophisticated forms, it was more a matter of holding breath than of moving forward significantly. Then the explosion happened, and there was a huge increase in life forms and in complexity. This is written in stone, as one might say. Well known are those one or two parts of the world, notably the Burgess Shale in the Rockies of western Canada, where there are Cambrian deposits of fossils that show not only skeletons but also the very softest of organs and bodies (Gould 1989; Conway Morris 1998). At first, the organisms they represent were thought to be very strange indeed, but detailed and continued study is starting to show how they relate to succeeding forms, sometimes (as expected) seeming to be the ancestors of forms that have diverged in dramatic ways. Included in these deposits are vertebrates and their immediate relatives and ancestors.

The reason for the Cambrian explosion is still mysterious. This does not mean that the time has now come to reach for miracles. There are many tantalizing clues, and causal hypotheses are getting increasingly sophisticated and plausible with the addition of new facts and techniques.

The answer probably lies in the confluence of several factors – biological, environmental, ecological. We are already prepared for the most significant biological factor. In Chapter 2, we saw that sophisticated animals – the bilaterian organisms (organisms that are symmetrical like us humans) – share a basic underlying genetic mechanism that orders and structures the production of the parts of the body (Carroll, Grenier, and Weatherbee 2001). These are the so called Hox genes that code for the proteins involved in the building of the organism – eyes where eyes should be, legs where legs should be, and so forth (see Figure 2.5). The Hox genes link humans and fruitflies, and show that we have a common ancestor. When did that common ancestor live? Techniques to be discussed shortly suggest that it was before the beginning of the Cambrian – at least a hundred million years before, and perhaps as much as a billion or more years before (Knoll and Carroll 1999).

In other words, for a very long time – however you calculate things – those genes were just sitting there without really doing very much. At least – let us rephrase this – they were doing something, namely, that which comes naturally. They were producing the bilaterians in the fashion that they do today, but for some reason the bilaterians could not really get off the ground, literally or metaphorically. Some have thought that it was all a matter of producing skeletons, but although skeletons may well have been useful once the Cambrian got going, it cannot be precisely this, because there are pre-Cambrian skeletal forms – not much, and not like those of today, but something. The point is that (one of the most important parts of) the genetic mechanism was in place to produce fantastically varied and complex organisms. And yet, until the Cambrian, it did not really get working to its full extent.

Turn now to ecology. Probably, as the explosion began, the first forms (of any kind) were pretty primitive – pretty wonky by today's standards. The now-vanished, nonbilaterian, pre-Cambrian forms, whatever they were, would have been good enough to keep the bilaterians down. Then something happened to wipe out the nonbilaterians or to render them less effective. The obvious and very well-established analogy is with the dinosaurs and the mammals. Both appeared about 200 million years ago, but for over 150 million years the dinos ruled the world and the mammals kept themselves inconspicuous: little nocturnal creatures that stayed underground or in the bushes. Then, when the giant reptiles went, the mammals could themselves take over and become the dominant form of large animal.

So, turning now to environment, given that we now know that the dinosaurs were wiped out by a mega-catastrophe – a comet hitting the earth about 65 million years ago and causing a nuclear winter through which the dinosaurs could not live (but the mammals, keeping warm and being small, could) – was there something comparable at the beginning of the Cambrian, or just before? The answer seems to be yes (Hoffman et al. 1998). There were massive ice ages making it very hard (although not impossible) for organisms to survive. Why these came on is a matter of some debate, but the moving of the continents was probably involved, and once enough ice was produced, the heat from the sun was simply reflected back into space, and then there was even more freezing. However, these times were unstable. After enough material was released by volcanoes to make for a gaseous blanket of an appropriate kind (carbon dioxide), warming started and caused the ice sheets to melt.

And – one final roll of the barrel – bingo! The freezing cut off the production of carbon dioxide by living organisms – no respiration was possible. This meant that oxygen was not being absorbed. Enough sunlight got through to keep up photosynthesis. Hence, oxygen levels started rising significantly. But we know that increased levels of oxygen are absolutely essential for the production of complex bilaterians. Hence, as the ice ages ended, the way was opened for the production of the denizens of the Cambrian. "Microscopic animals with scant oxygen requirements could have plied Proterozoic seas long before the Ediacarian epoch. Only with a latest Proterozoic rise in oxygen, however, did macroscopic (and, hence, easily fossilizable) animals become possible" (Knoll 2003, 218).

And with this, life as we know it was under way. Is the story just given absolutely and completely true, beyond reasonable doubt? Almost certainly not. There will be revisions and substitutions, with other ideas coming to the fore. But the scenario just sketched has some solid evidence behind it – it uses ideas and findings that we simply did not have one or two decades ago – and it points the way to similar and other avenues of research. Leave things at that.

Life Under Way

Moving forward now, and starting to hitch some absolute dates to the Victorians' picture, the move to land probably occurred even before the Cambrian, but it was the Devonian that saw the major invasions of land plants and animals, with the subsequent Carboniferous being the time of the great forests (now converted to oil and gas and coal). Insects first appeared in the Devonian, took to the air during the Carboniferous, and then later became vertebrates. As just noted, mammals arose about 200 million years ago, but remained small and fairly insignificant until about 65 mya, when the dinosaurs disappeared and the so-called Age of Mammals started. Birds have been around since about 150 mya (the late Jurassic). As is well known, general thinking today is that they evolved directly from the dinosaurs. In the plant world, the angiosperms – the flowering plants – arrived about 130 million years ago (the Lower Cretaceous). These are organisms that require pollination, and although this can be done by the wind, insects play a major role. It goes almost without saying that the arrival of insects that use vegetable matter for food was particularly important. "The evolution of the herbivore guilds was strongly connected with the evolution of the angiosperms and a global increase in vegetation complexity" (Dmitriev and Ponomarenko 2002, 376).

Like the unraveling of the Precambrian, adding absolute dates was not so much a question of radical revision of the earlier known fossil record as it was one of augmentation. What then can we say about the quality of the record, particularly from the Cambrian on? One can answer this question in a number of ways. Have we reason to think that we now have a good reading of what happened and that new discoveries are going to fill in gaps rather than call for drastic revisions? The answer to this question seems to be an emphatic yes (Benton 2006). As the hotel advertising assures us, the best surprise is no surprise. This has been the story for the past 150 years. Huge amounts of new material have been uncovered, starting with all of the magnificent finds in the second half of the nineteenth century in the American West. Thanks to such rivals as Edward Drinker Cope and Othniel Charles Marsh, dinosaurs of all shapes and sizes, mammals like nothing ever seen before or since, and many more weird and wonderful forms were unearthed (Shor 1974). And the important point is that, magnificent though these finds surely are, they all fit in. No one is worrying about a discovery of reptiles in the Cambrian, for instance. There were none!

Of course, one might interject that the conclusion that we have merely filled in the already-known record makes a significant assumption. We are assuming that there is no factor seriously distorting our reading of the record, and that this has been the case all the way from Owen to the present. If such factors exist, then the best we can say is that those parts we know are getting ever-more-carefully uncovered and understood. But occasional claims to the contrary, such distorting factors seem not to exist. On the one hand, there is no real reason to expect them – within limits, everything does go smoothly from one form to another (expected) form. If there are major unknowns, then they must be doing something really remarkable, like going off at tangents and leaving no record, or going off and then reversing, again leaving no record – and the more remarkable, the less probable. On the other hand, across the globe one can correlate strata, and there simply does not seem to be evidence of major times when the record was not being laid down – certainly not since the Cambrian. If there are gaps, where are they and when?

Another way to answer the question is to focus on particular links and paths. Does one have transition fossils – missing links – between major groups of organisms? Can one spell out in some detail transitions from one form to another very different one? Without pretending that one can always answer these questions positively, without indeed pretending that one can usually answer these questions positively, the answers are that very significant links are known, and detailed transitions can be given. Other than humans – the topic of a later chapter – the most famous link of them all is that (mentioned earlier) between the reptiles and the birds. As it happens, this fossil, although not properly classified, was known even before the *Origin*, but not until shortly thereafter was a complete specimen unturned and its true significance recognized. I refer to the transitional reptile-bird Archaeopteryx, uncovered in slate mines in Germany (Benton 1990). It has feathers, so is as bird-like in its all-defining major respect as it is possible for a brute to be. But like a reptile (and unlike a bird), it has teeth; it has a reptilian-sized brain; it has separate (rather than fused) digits; it has a reptile-like breast bone, and much more. You could not have a better transitional form if you had designed it yourself (Figure 4.4). It is true that Archaeopteryx was probably not the actual link between the reptiles and today's living birds – it was off on a side branch. And there are other forms that now have stronger claims in this respect. But this is not to diminish its worth, for (given the branching nature of evolution) one expects there to be a divergence of the earliest form or forms and that therefore most subsequent forms will go extinct. Indeed, the wonder would be rather if Archaeopteryx were the actual ancestor.

FIGURE 4.4. Comparison of pigeon (on left) and Archaeopteryx (on right). From E. H. Colbert, *The Evolution of the Vertebrates: A History of Backboned Animals through Time*, 2nd ed. (New York: Wiley, 1969). Reprinted with permission of John Wiley and Sons, Inc.

The point is that it is right on the divide, as one expects if evolution be true.

What about detailed transitions? These are common also, and at a broad level can cover some of the most important of the changes. There are, for instance, so many examples of reptilian-like mammals or mammalian-type reptiles that it is difficult to know where to draw the actual transition line. At a more detailed level, one of the recent (and ongoing) triumphs is that of the transition from land mammals back to the sea – the evolution of the whales and so forth. No one thinks that whales are the original mammalian form. Rather, the original mammals were land animals, and then for various reasons some started to move back toward the sea – first otter-like animals, and then seal-like, and then finally whale-like forms. (Actually, one first had hippopotamus-like animals on land, and then the move was made to more and more aquatic forms.) Now we have some fantastic fossils, starting about fifty million years ago with mammals that

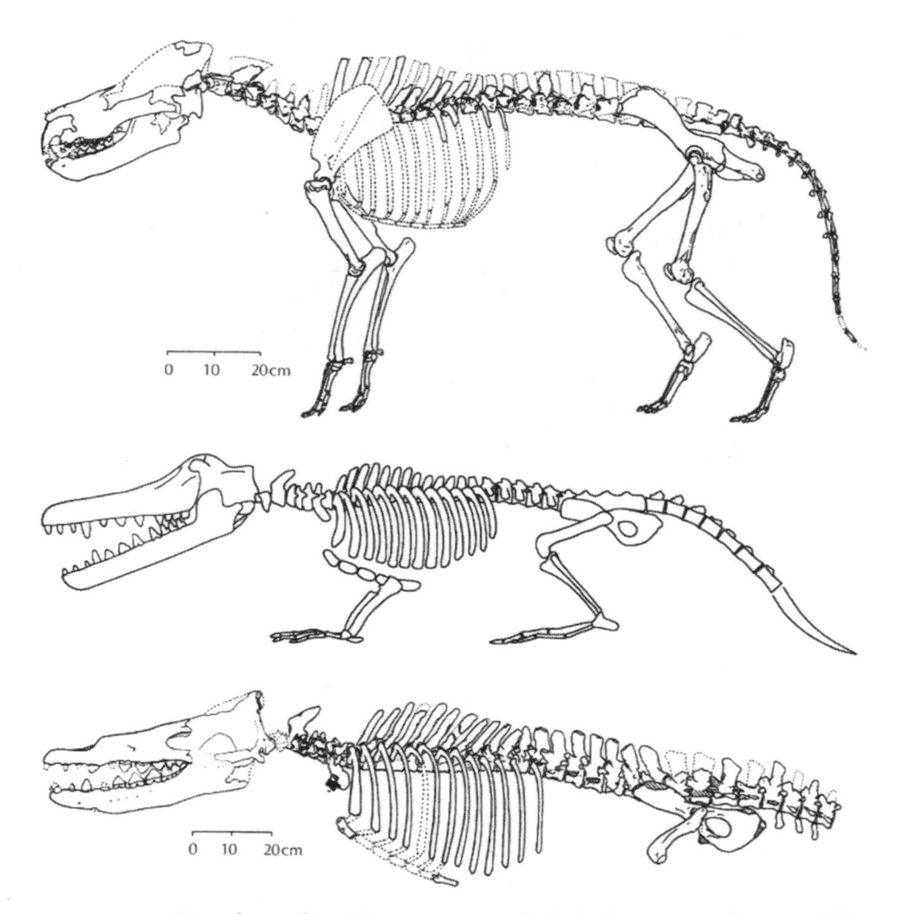

FIGURE 4.5. Top, the earliest Eocene mesonychid, *Pachyaena ossifraga*. Middle, the earliest middle Eocene whale, *Ambulocetus*. Bottom, the earliest middle Eocene whale, *Rodhocetus*. It is thought that the whales descended from the mesonychids or like animals. *Pachyaena* has features associated with a semiaquatic lifestyle. From Philip Gingerich et al., "New Whale from the Eocene of Pakistan and the origin of Cetacean swimming," *Nature* 368 (1994): 844–7. Reprinted by permission.

were land-based but clearly marine-using; moving then to animals that were essentially marine-based but still had semifunctional hind limbs (perhaps used for things like grasping in copulation rather than for swimming, and certainly not for walking); and moving right up to the present, where we have mammals that simply have no real limbs at all and are completely marine-based. You could not ask for a better picture (Carroll 1997; Dawkins 2004) (Figure 4.5).

Patterns

Ask now a somewhat different question or set of questions. Does the fossil record show any definite pattern or patterns? Leave for a later chapter the overarching question about whether the record in any sense shows improvement – progress – and ask rather about trends and themes and patterns. We have already been introduced to the notion of mass extinction – the putative extinction(s) just before the Cambrian, and then the much celebrated (and indisputable) extinction 65 million years ago, at the K/T boundary. This latter was brought on by a foreign object – a comet or the like – hitting the Earth, just off Mexico (Alvarez et al. 1980). There have been several other extinctions spread over the past 500 million years, between the Cambrian and the Tertiary, some with effects surely as great as either of these – 90 percent of animals and plants being wiped out. These extinctions may also have been caused by extraterrestrial factors, but some may have had a more local origin, namely, the changes in habitat (around sea shores, for instance) brought on by the changing and moving of the continents, in turn caused by the shifting plates on which they ride.

The extinctions were significant. Without the first, we would probably have no life as we now know it. Without the last, I would almost certainly not be writing this book. In other respects, there is debate about their lasting significance, especially since the beginning of the Cambrian. Some suggest that, in the long run, the general patterns of the record do not change much. There are two basic competing hypotheses used to structure and explain the record (Figure 4.6). The first – the *logistic* model – argues that we get waves of a kind of sigmoidal growth followed by equilibrium (Ruse 1999b). This hypothesis was the brainchild of the late John J. Sepkoski (1978, 1979, 1984), who argued that organisms make a breakthrough into a new ecological niche, then start to expand in numbers – not so much individual numbers, but diversity numbers, as represented by numbers of taxa, like genera or families. At some point, the available space (ecological rather than purely physical) starts to fill up, and so the growth evens off, producing a familiar S-shaped curve. Eventually a kind of equilibrium is achieved and maintained, until again a group breaks through into a new space, and the growth starts all over again. The old forms of organisms persist (and keep evolving) but generally suffer some decline in numbers, so that the overall effect is a kind of layering of new upon old. Sepkoski argued that this equilibrium type of situation holds in

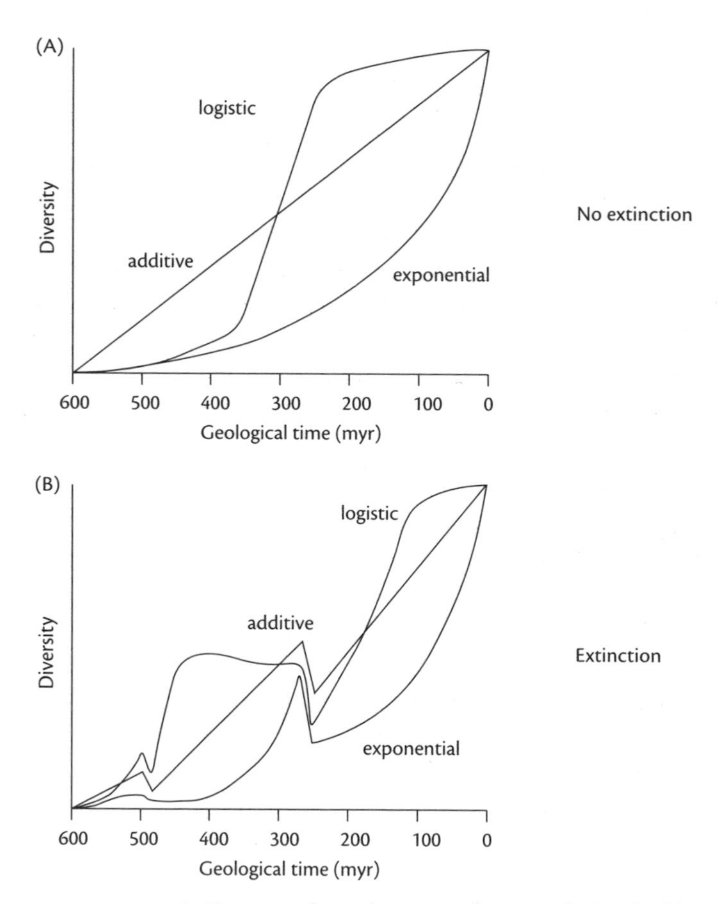

FIGURE 4.6. Diversity through time as shown at the level of families.

plants as well as in animals, and that indeed we might today be seeing a new breakthrough in the plant world (Figure 4.7).

The chief rival to the logistic model is the *exponential* model (Benton 1987, 1997, 2006). This model relies on the assumption that, if one takes them over reasonably long periods of time, rates of speciation and of extinction remain roughly constant. This leads to an exponential rate of growth, and the main thesis is that really there need be no limit to the growth. In other words, rather than just finding niches existing independently that they then occupy, organisms themselves create the niches,

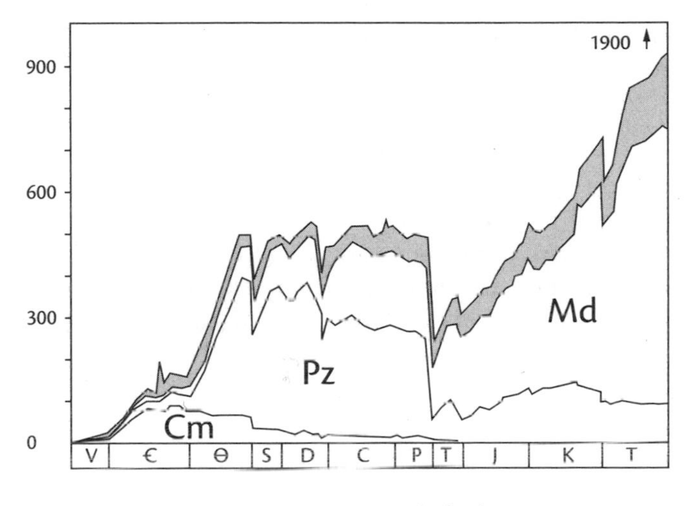

FIGURE 4.7. From J. J. Sepkoski, Jr., "A kinetic model of Phanerozoic taxonomic diversity. III. Post-Paleozoic families and mass extinctions," *Paleobiology* 10 (2): 246–67. Courtesy of the Paleontological Society.

and thus there are always more available. Hence, at some basic level, this model rejects equilibrium thinking and argues that life is always on some kind of move upward. Supporters of this exponential model recognize that not every kind of organism keeps expanding as it would predict. But whereas an equilibrium model is more likely to attribute extinction to failure in competition with other organisms – someone has got to give – the exponential model gives a larger role to external extinction events clearing things out, irrespective of the biology of the situation.

The jury is still out on these two hypotheses. Perhaps the truth lies in a combination of the two. It has been suggested that the sea might be more stable and lead to greater restriction of possibilities. Hence an equilibrium-based model like the logistic model might be appropriate in that environment. Land offers more opportunities, and hence an exponential or expansionist model might be more appropriate. More work is needed, although before the reader rushes to the conclusion that we have rival models that are never going to be resolved or reconciled, note that both hypotheses are fairly new – no more than two or three decades old – and that their formulation (let alone resolution) was really not possible before the development of computers capable of digesting and processing huge amounts of information. In respects, we are only beginning to get into questions of this nature.

Cladism

The fossils are not our only entrance into the past. The pre- and nonevolutionists of the eighteenth and nineteenth centuries knew that morphology and embryology can tell us things, even when there is no direct evidence in the record. Assuming that organisms were created (naturally or supernaturally) in some kind of sequential and branching fashion, then the ways that they are now can tell us much about the ways that they were then. Again, the *Naturphilosophen* led the way, seeing life's history as something paralleling the development of organisms (Russell 1916). One such thinker was Louis Agassiz – he of the "prophetic types" mentioned earlier – the great ichthyologist who moved from Switzerland to Harvard. Although never an evolutionist, he was very much into parallels between life history and embryological development; and indeed, he saw a threefold parallelism that included living beings today. "One may consider it as henceforth proved that the embryo of the fish during its numerous families, and the type of fish in its planetary history, exhibit analogous phases through which one may follow the same creative thought like a guiding thread in the study of the connection between organized beings" (Agassiz 1885, 1, 369–70). All of this led very easily to an evolutionary interpretation, especially at the hands of Darwin's most enthusiastic German supporter, Ernst Haeckel (1866). Trees of life proliferated, and embryological analogies ran rife, especially thanks to the so-called biogenetic law, the evolutionary equivalent of the organism–history analogy: "Ontogeny recapitulates phylogeny."

The biogenetic law proved to have many, many exceptions, so many that it was often rejected entirely (Gould 1977). By the end of the nineteenth century, ambitious young biologists were using its overemployment as a reason (with some justification) to turn from evolutionary studies to other, more fertile fields. But this does not mean that embryology as such is without value in ferreting out life's history. Tying it in with the branching nature of the fossil record, the Estonian nobleman Karl Ernst von Baer had argued that we see a move embryologically away from the basic form toward various refined forms, not all on the same branch. "The general features of a large group of animals appear earlier in the embryo than the special features" (Ridley 1986, 66). In other words, even though there is no direct recapitulation – the embryos of an organism are like the embryos of other organisms, not like their adults (except incidentally, if the other organisms do not change much from embryos to adults) – one can use embryos as clues to history. One can even do this at times when, because the adults are very different, it would be difficult or impossible to make comparisons on the adult forms.

Basically, this approach to the past – morphology augmented by embryology – turns everything into questions of similarity and difference. Organisms that look alike are going to be considered more closely connected – for the nonevolutionist, because they were created (naturally or nonnaturally) at an earlier time, perhaps from a shared template (*Bauplan*); for the evolutionist, because they are descended from a shared ancestor, which means they are more closely related than mere strangers. Organisms that do not look alike are going to be considered less closely connected – for the evolutionist, less closely related. In both cases, for the evolutionist, the assumption is that looks reveal actual, ancestor–descendent links. I look more like my brother than either of us look like a total stranger; hence, the inference is that my brother and I are related more closely than we are to the stranger. But there must be more than this. "Looking alike" is a weasel term – does one record a shared count of a thousand long hairs as one feature or a thousand? And what about changes that occur just because of shared circumstances or functions? Does one put bats with fruitflies because both have wings and cows do not, or does one put bats with cows because both suckle their young and fruitflies do not? From a history-seeking perspective, the trick is going to be to capitalize on features that reflect shared ancestry (or the equivalent for the nonevolutionist) – the "homologies" of which we have spoken earlier – and to avoid those that simply reflect coming together in evolution (or nonevolutionary time), perhaps because of a common function – so-called "analogies" (or "homoplasies," in modern language).

Many of these insights have been formalized recently (and much use has been made of the power of computers), as systematists devise classifications that reflect putative ancestry. This approach is called "cladism" or, in older literature, "phylogenetic systematics" (Hull 1988; Ridley 1986; Sterelny and Griffiths 1999). The key is the identification of shared derived characteristics – that is to say, characteristics that were not possessed by the ancestors of the groups but that have been acquired along the way by some

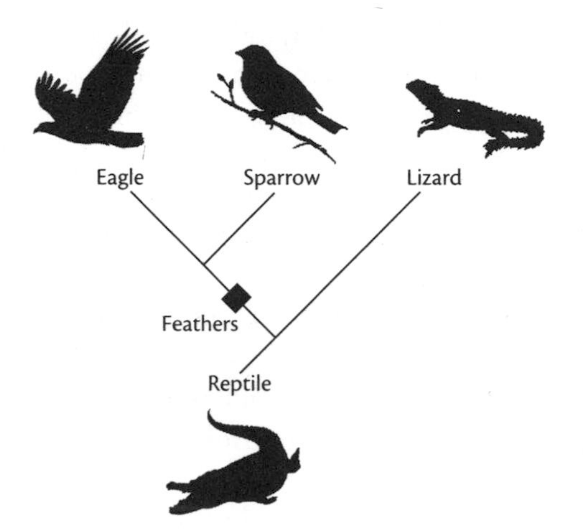

FIGURE 4.8. Cladogram showing that feathers needed to evolve only once.

(but not all) of the descendents. For the apes, for instance, large brains would be shared derived characteristics, not possessed by the early mammals or by many other mammals today. Unfortunately, perhaps because the founder of cladism was a German (Willi Hennig in the 1950s), and perhaps also because systematists are insecure about their status, fearing that others judge them as little more than stamp collectors and not true causal scientists (there is some truth to this fear), the language of cladism steers somewhere between the baroque and the barbarous. Shared derived characteristics, for instance, are known as "synapomorphies." We can ignore the Teutonic tongue-twisters. The ideas are basically simple – and (to be fair) powerful. Suppose we have three organisms – let us say, an eagle, a sparrow, and a lizard. How do we group these? Fairly obviously, we are going to put eagle with sparrow and leave the lizard off to one side. If challenged, we would say that lizards, with scales, are still in the ancestral condition, and that being feathered is a feature gained by birds along the way. Our supposed phylogeny would look like Figure 4.8. Actually, this is not strictly a phylogeny, but what is known as a "cladogram." Phylogenies may be what we hope to get by the time we are finished, but at this point we are, at best, only part way there. There is no absolute time fixed to anything. Also, actual changes are treated somewhat artifactually. It is assumed that all changes involve branching (into two lines, no more) and that the changes occur at the time of branching. In other words, feathers are the feature of everything on the left, after the first branch. (Hence

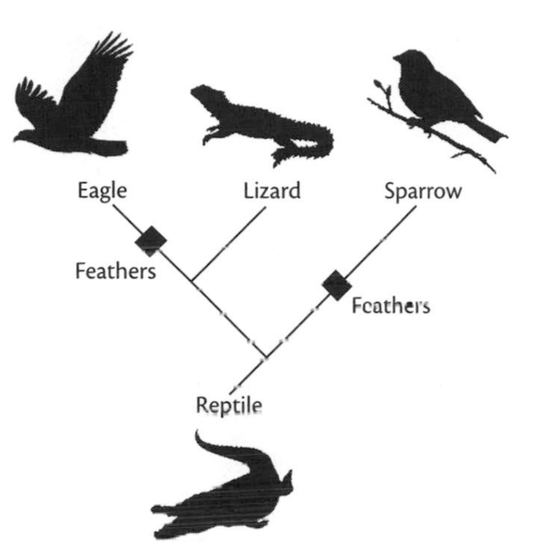

FIGURE 4.9. Alternative cladogram showing that feathers needed to evolve twice.

the name "cladism," since a clade is all of a group of organisms from one ancestral group.)

If we had the fossil record, we could check things out. Generally we do not have the record. So now the question comes, why the cladogram in Figure 4.8? Why not Figure 4.9, where the lizards broke off from the eagles? Here cladists usually appeal to simplicity, what they call parsimony (Sober 1988). Figure 4.8 shows a simpler pattern than Figure 4.9. In Figure 4.8, you need only one move to feathers, at the point of the first branching. In Figure 4.9, you need two moves to feathers, at both points of branching. (There are other, even more complex possibilities, which involve an initial branching to feathers and then a later move back to scales for lizards.) Important in the cladistic technique are organisms known as "outgroups." Suppose you are faced with horses, zebras, and cows. What are you going to ignore? Are you going to worry that backbones are things that came and went and perhaps came again? You compare these groups to something clearly not so closely related, like a snake, and find that snakes also have backbones, and so this is almost certainly a feature that you can ignore. It is not a derived feature, but ancestral. (You cannot just assume that the outgroup is not so closely related – you need independent evidence, like pertinent fossils.) Then, with the irrelevant features identified, you are off and running, looking at shared features (like one digit in horses and zebras versus cloven hooves in cows), and so on and so forth. It is highly unlikely that one feature will give all of the answers, so you run the changes on many features, generating multiple cladograms, and then you put them together in the way that requires the least number of moves or changes. Particularly in this day of computers and of molecular biology (which reveals many easily quantifiable and recordable features), this technique has proven very powerful, giving new trees of life that are more detailed than anyone a few years ago dared to think and showing relationships that few supposed even in their wildest dreams. Return again to the Precambrian era (Knoll 2003; Freeman and Herron 2004). For all of the important finds, the fossil record comparatively remains dreadfully sparse. Comparison between today's molecules throws light where hitherto there had been total darkness. One amazing finding is that the apparently sacrosanct division of organisms into eukaryotes and prokaryotes – us versus them – turns out not to be quite as fundamental as we had thought. It is a very important division, so nothing said thus far in this chapter is negated or really even belittled, but it is not the unique division in the history of life. Our kingdoms (animals and plants) are small branches of one of three big groupings, as can be seen from a recent all-inclusive phylogeny (Dawkins 2004, 461) (Figure 4.10). Eukaryotes, those cells with nuclei, fall into a mega-group along with other such organisms as fungi and slime moulds. Then there are the prokaryotes, divided into the Archaea and the Eubacteria. The former contain the kinds of bacteria that can survive in extreme conditions, such as superheated underwater vents and highly saline waters. The latter also contain bacteria that are able to withstand high temperatures, as well as other primitive organisms, including those needed for photosynthesis. It has not gone unnoted that these high-temperature organisms might be the most primitive of all, and hence were present at the very beginning of life.

It has also not gone unnoted that this picture of a life as a tree might need modification – outright rejection, according to some. Given the truth of Lynn Margulis's hypothesis about the origin of eukaryotes, already we have been pushed toward more of a network-like picture, with branches crossing over to other branches, than that of a conventional tree. If (as mentioned earlier) lateral gene transfer between branches (thanks to viruses and the like) is common, then the old picture of separate lines reaching upward is more than misleading. Probably – at least as we understand things at this stage – there are as yet no grounds for radical rejection or revision. Evidence thus far is that lateral gene transfer is more a complication of a basic, tree-like picture than something that demands a

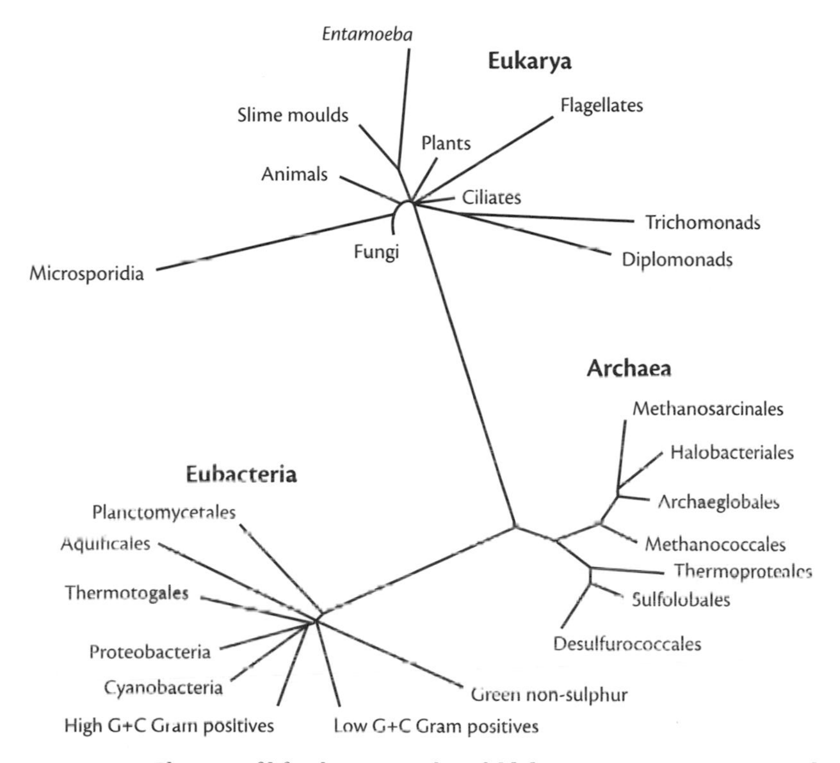

FIGURE 4.10. The tree of life, showing its threefold division. From R. Dawkins, *The Ancestor's Tale*, with permission. Adapted from S. Gribaldo and H. Philippe, "Ancient phylogentic relationships," *Theoretical Population Biology* 61 (2002): 391–408.

totally new approach. It could well be that the really fundamental genes – those that carry the information for the building of the cell – do not get transferred, and that it is only the secondary genes – those that alter the operations of the cell – that can move over. You are not going to change a cow into a horse, but you might make a bacterium resistant to heavy metals. (Note, incidentally, how lateral gene transfer undermines the claim that genetically modified foods are "unnatural." Nature is often inserting alien genes. This does not, of course, imply everything we do is safe!)

Problems?

Two philosophically oriented questions need answering. First, what about the assumption of simplicity? This raises general issues – many think that simplicity should have no role (except perhaps a reluctant role) in science. It is true that often there are pragmatic reasons why scientists opt for the simpler solution – "Keep It Simple, Stupid!" – but that is a reflection of our limitations rather than a truth about the world. It is also true that if one thinks that there is a Creator behind everything, then there may be a presumption that simplicity is more than just a mark of human inadequacy. Often simplicity is linked to a kind of elegance, and this is a very powerful force in science – "Something that pretty just has to be true!" – and presumably the Creator is no less keen on beauty than we. But if you drop God, or at least if you take God out of science in the fashion of the naturalist, then simplicity again rears its ugly head. Why should we accept the arguments of the cladists, given that assumptions of simplicity are at the heart of their methods?

There are two responses that one can make here. First, there is a general defense of simplicity as such. Almost everyone now recognizes that the idea of science being no more than a faithful representation of reality – a kind of police pathologist's photo – is chimerical (Ruse 1999b). Science involves a huge amount of interpretation and creative understanding – in many respects, it is more like a Van Gogh painting, with all of the interpretation that that implies, than a straight photo (assuming that photos, even police pathologist photos, do not involve interpretation). So assumptions of simplicity (clearly involving interpretation) don't merit any special comment. Moreover, because you have made an assumption of simplicity, this does not mean that your thinking is now forever frozen into one pattern. Start simple, but then get more complex as the facts dictate. This is as true of cladism as of the rest of science. (There will be much more on science and its interpretations in a later chapter.)

The second, more specific point of response is that cladistic assumptions about simplicity are not just reflections of inadequacy, pushed onto interpretations of the real world. All that we know about evolution – including what we know about causes (to be discussed in the next two chapters) – suggests that similarities between organisms are more likely to be reflective of homology than of convergence and analogy. You start with homology but have to work to (that is, evolve to) analogy. So going with the simplest option is an evidentially supported strategy – there are good empirical and causal reasons for going this way. It might prove to be wrong, but it is not just a reflection of the intellectual limitations of cladists. (Some would say that similar comments can also be made about the cladists' general assumptions that change takes place at points of branching, and that such branchings involve bifurcation rather than multiple speciation. We shall have more to say on some of these matters in later chapters.)

A second query might seem more devastating, although in fact it can be cleared up quickly. Why should we think that an approach that relies not on the fossils but on morphological similarities and differences is truly evolutionary? It is notorious that the coming of evolution – of the Darwinian revolution – made little difference in actual practice to the work of biologists. The Watson–Crick DNA model turned genetics right around. Evolution had no such effect on the working biologist, save for those few actually working on Darwin's mechanism of natural selection. Reading the work of biologists in the middle and late nineteenth century, you often cannot tell whether or not someone is an evolutionist (Ruse 2005b). Idealist morphology and evolutionary morphology blended right into each other. In both cases, one is really just working out formal patterns based on extant characteristics. In which case, why pretend that any of this gives you genuine insights into the past?

This is a query that persists down to the present, for influential cladists have asked precisely the same question.

> As the theory of cladistics has been developed, it has been realized that more and more of the evolutionary framework is inessential and may be dropped. . . . [Cladist Norman] Platnick refers to the new theory as "transformed cladistics" and the transformation is away from dependence on evolutionary theory. Indeed, Gareth Nelson, who is chiefly responsible for the transformation, put it like this in a letter to me this summer: "In a way, I think we are rediscovering preevolutionary systematics, or if not rediscovering it, fleshing it out." (Ridley 1986, 87, quoting the English taxonomist Colin Patterson)

There is a ready response to be made here. Even if we agree that much of the classificatory procedure is not evolutionary – although today, surely, in many branches (especially those with some fossil record) it is going to be difficult to separate out what is not evolutionary from what is – this does not mean that classification has no relevance to evolution, or that evolution has no relevance to classification. As soon as you start to ask causal questions (and I have just suggested that even if you do not ask them, such questions lurk behind vital cladistic assumptions about simplicity), you come up against evolution. At least you do if, like any modern scientist, you make a commitment to a naturalistic approach to understanding. From your comparisons, you may get only a formal pattern – the picture just given of life's three great divisions may represent just such a pattern – but what makes sense of the pattern is the fact of evolution. Why do you have the pattern? Because of evolution! What does the pattern tell you! The path of evolution! Your procedures may be pre-Darwinian. Your understandings are post-Darwinian. Until someone comes up with an alternative naturalistic explanation, that is all that there is to be said on the subject.

Molecular Clocks

There is a third way of digging out phylogenies, related in a fashion to the second, but best treated as a method on its own. One major difficulty with cladism is that, although it may tell you about the relative relationships of one organism to another, and may point you back in time to where there were points of divergence, it is of no direct help in putting absolute dates on things. Plants divided from animals back at some point, but when? Humans sheared off from other animals in the past, but when? Here it seems that the fossil evidence is definitive – use radioactive clocks to determine major events in the record. But this means in turn that where there is no fossil evidence – especially in the pre-Cambrian, not to mention for organisms like bacteria, which hardly ever leave traces – you are stuck with only relative dates. A is older than B, but not as old as C. We may agree that there was some shared ancestor of humans and fruitflies that had Hox genes, but how can we know that this organism flourished millions of years before the Cambrian?

Fortunately, there is a way in which you can use comparisons of present data from organisms to estimate absolute and not just relative dates. The theory is simple, even though in practice it can get quite complex. Forget about what goes on at the physical (including behavioral) level of organisms. Concentrate on the molecules, proteins or, more recently, the DNA molecules. We know that there is lots of redundancy at this level – for instance, the cell uses only twenty amino acids to build proteins (the building blocks and enzymes of the cell), and yet the genetic code has sixty-four possible combinations. (Four bases, grouped in threes, to give 4 to the power 3, equals 64.) So, many third places in the code are interchangeable with many other bases. And then again, we now know that much of the genome is filled with stuff that does nothing – junk DNA. So, even if many places in the DNA molecule are filled with bases that are working, there are many places were the bases are not working, and even those places that have workers can have alternatives. But mutation goes on all of the time, affecting nonworking slots as well as working slots (including duplicate slots). Since nonworkers and duplicates (inasmuch as they are duplicates) are going to make no difference to what happens up above – at the physical level – then, whatever the cause of evolution, one expects that the molecules at these slots will vary more or less randomly with respects to upper-level causes. What will affect their variations will be their rates of mutation.

This fact at once opens up the possibility of assessing relationships (in a way akin to cladism), but more than this, of assessing the absolute times of past events (in a way akin to the absolute-time assessments of fossils) (Hillis, Huelsenbeck, and Cunningham 1994; Hillis, Moritz, and Mable 1996). To do this, you need some markers drawn from the fossil record. But suppose, for example, that you know (from the fossils and radioactive clocks) that some event of divergence occurred ten million years ago. By checking appropriate molecules of the divergent organisms, one can calibrate how far apart they have "drifted" during that time. Build in now the true assumption that the rate of increase of random variation (unlike variation under the control of evolutionary causes) is going to be steady – in the long run, that is, rather like radioactive decay – and you have a measure of other events in the record. Suppose you have other organisms in the same group, and suppose they are twice as different as those that you can calibrate as dividing ten million years ago. Then (even without fossil evidence) you can say that they broke apart twenty million years ago.

The "molecular clock" has proven a very powerful tool in exploring the past, and in particular is the means by which almost all of the pre-Cambrian dates are calculated. Like anything in biology, there are complications and distorting factors that need to be taken into account. Generally, for instance, one needs to take note of reproductive rates. Since most mutations occur at times of reproduction, the expectation is that fast reproducers will show more change than slow reproducers. This is not always entirely true, but it is something that must be noted and, if necessary, accounted for. Likewise, one finds that some molecules are entirely independent of what goes on at higher levels, and others are only slightly connected. Then again, redundant molecules might not be entirely random in their placement. Although evolutionary factors may not impinge on them, it could be that straightforward physico-chemical causes are liable to make some options more probable than others. But all of these are problems calling for refinement, rather than barriers calling for rejection. The clock has proven its worth in going back in time.

Consilience

One final point and this chapter can end. If we do have these various ways of examining the past, do we find (as we should expect) that they bring us independently to the same answers? Do we have a consilience, much increasing our confidence that we are truly finding out the past path of life? Are our phylogenies reliable? In truth, of course, the methods are not entirely independent – we have just seen how the molecular clock initially needs fossil evidence to calibrate it. But the methods do employ very different approaches, and coincidence is important. And it is gratifying to be able to say that such coincidence is very much the rule rather than the exception (Benton and Storrs 1994; Wray 2001). Cladistic analyses done independently of the fossil record do then fit with the record, and molecular estimates of time likewise fit with the record and with divisions made cladistically. Not always, expectedly – especially not at first. Most famously, traditional divisions of humans from other animals put the date back to about twenty million years ago. Then along came the molecular biologists, suggesting that the division happened only five million years ago (Sarich and Wilson 1967). But initial disconsolation and disbelief (not to mention irritation that people from another field were the ones to put things right) rapidly gave way to recognition that tradition was wrong and the molecules right. Today, no one doubts that the division occurred only about five or six million years ago. (More on this in a later chapter.)

There is a consilience. The methods of finding out the paths of evolution agree. Gish is wrong. *Evolution! The fossils (and everything else) say Yes!*

CHAPTER FIVE

The Cause of Evolution

The Synthetic Theory of Evolution is dead.

S. J. Gould (1980a)

Stephen Jay Gould – Harvard professor, paleontologist, agnostic Jew, choral singer, baseball fanatic – is dead. He died in 2002, at the age of sixty, having twenty years earlier bravely fought and conquered a vile form of stomach cancer, and then for the next two decades having lived and shared a full and rich life, until he was finally felled. He was best known as a brilliant popular-science writer, for twenty-five years without fail penning a column, "This View of Life," in the monthly magazine *Natural History*. But he was at heart a professional scientist, promoting his own discipline and forever attacking what he felt was a false orthodoxy – a Darwinian view of life that promoted and focused on one mechanism, natural selection. He thought that this led to a distorted view of the evolutionary process and that it was in fact – whatever the talk – truly not the way that real scientists behaved and thought. Hence the notorious claim that heads this chapter, that neo-Darwinism or (as it is known in America) the synthetic theory – that blend of selectionism and modern genetics – is an exhausted theory or paradigm.

In this chapter, we shall look at the positive case to be made for selectionist thinking. In the next chapter, we shall look at some of the criticisms. Our topic simply is the adequacy of Darwin's mechanism: is natural selection the chief cause of evolutionary change? Do we have a differential reproduction of organisms brought on by a struggle for existence, and, as a consequence of such a process, is it the case that organisms are adapted to their environments? Do organisms show design-like features?

Population Genetics

Mendel's work was rediscovered at the beginning of the twentieth century. Almost as soon as this happened, a number of mathematically gifted thinkers began extending genetics to group situations. This theory of "population genetics" – which came to real fruition around 1930 – is the foundation of modern evolutionary thinking and the place where today we first encounter natural selection (Provine 1971). As we have seen, today the Mendelian gene is identified with lengths of DNA, but initially (as well as historically, obviously) there is no need to think in molecular terms. The woodworker shaping a piece of oak with plane and chisel does not need to think in terms of molecules. However, when he is doing some things – for instance, trying to bend the oak under pressure and hoping to have it stay in shape – knowledge of the molecules might become pertinent. So likewise in population genetics.

Unlike Darwin's theory of the *Origin*, which introduced natural selection head on, modern theory starts elsewhere, and only secondarily introduces selection. This does not mean that the importance of selection is minimized – it is more a question of structure and presentation. The best model or analogy for what is happening is given by Newtonian mechanics. The key force there is gravitational attraction. It is this that explains everything, like the planets going in ellipses around the sun and cannonballs here on earth describing parabolas. But one does not start with gravity. One begins with the three laws of motion. Most particularly, one has the first law, namely, that bodies continue in a straight line in uniform motion, or at rest, unless acted upon by a force. This is an equilibrium law. If nothing happens, then nothing happens. Its importance is that it gives you a background of stability that can then be used to describe what happens when something (namely a force) does happen. If there were not this background of stability, then you simply could not make sense of anything. Any event would be random.

We have the same situation in population genetics (Ruse 1973; Sober 1984). You have forces – most particularly, natural selection – but you must first have a background of stability. It is genetics that gives us this,

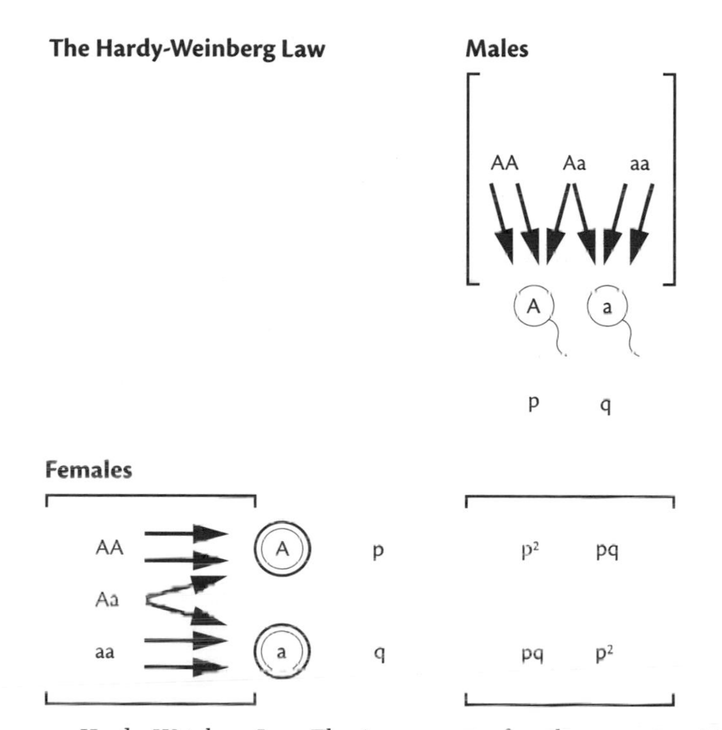

FIGURE 5.1. Hardy–Weinberg Law. The A genes come from homozygotes, AA, and from heterozygotes, Aa. Similarly for the a genes. The ration of A to a is p to q, where $p + q = 1$. Since there are p genes proportionately coming from sperm and p genes coming from ova, there will be p^2 AA homozygotes, and so forth. Since the new generation now has $2p^2 + 2pq$ type A genes, and $2pq + q^2$ type a genes, simplifying this again yields the ratio of p:q.

and Mendel's first law is the key, for this law tells you that genes get passed on unchanged from generation to generation, in a systematic way. What is needed is this law applied to large (effectively infinite) populations. This generalized law is named after the two men who found it – the English mathematician G. H. Hardy and the German Willhelm Weinberg (Figure 5.1). Understand the full import of the law: p^2 AA + $2pq$ Aa + q^2 aa. What it tells you is that you will always have the ratio p:q and that (after the first generation, and no matter what the initial distribution) you are going to get these distributions indefinitely, no matter how rare or common an allele may be. The rare one will never get lost, so long as there are no external forces, nor will the more common allele become more common. Things stay in equilibrium.

Change of gene frequencies when the homozygote AA and the herozygote Aa are fitter than the homozygote aa.

	Genotypes			Total Population
	AA	Aa	aa	
Initial zygote frequency	p^2	$2pq$	q^2	1
Fitness, w	1	1	$1-s$	
Zygote proportions after selection	p^2	$2pq$	$q^2(1-s)$	$1-sq^2$
Zygote frequencies after selection	$\frac{p^2}{1-sq^2}$	$\frac{2pq}{1-sq^2}$	$\frac{q^2(1-s)}{1-sq^2}$	1

One can now calculate the change of gene frequency (written as Δp) in the population. If the new frequency is X, then the change is X-p, and, remembering that each organism has two alleles, and that the AA homozygote has two A alleles and the Aa heterozygote has one allele, then

$$X = \frac{2p^2 + 2pq}{2(1 - sq^2)} = \frac{p(p+q)}{1 - sq^2} = \frac{p}{1 - sq^2}$$

Hence

$$\Delta p = X - p = \frac{p - p(1 - sq^2)}{1 - sq^2} = \frac{spq^2}{1 - sq^2}$$

Immediately one gets a fascinating and counter-intuitive result. If a gene, "a" in this example, is recessive, then it is going to be very difficult to eliminate it, even if you remove every visible case in each generation. If you start off with the gene being equally represented ($p = 0.5$), then – even if not one homozygote (aa) breeds – it will take a thousand generations to get "a" down to a ratio of one in a thousand. For slow breeding organisms like humans, this is going to take many years, and has obvious implications for programs ("eugenics") aimed at eliminating the genetically unfit.

FIGURE 5.2. Population genetics in action.

At this point evolutionists introduce disturbing factors – forces that are going to alter gene ratios. One set will involve immigration into and emigration out of the population. If, for instance, those homozygote carriers of A alleles are likely to vanish into another group, then this would affect ratios over time. Another set of forces could be mutation. If in each generation a proportion of a alleles mutate to A alleles, then this will clearly make a difference over time. And finally, and most importantly, there is selection. Suppose that having at least on A allele makes you fitter than an organism without such an allele. In other words, AA homozygotes and Aa heterozygotes are fitter than aa homozygotes. There will be changes over generations, and it is fairly easy to quantify these assumptions. The fitness, w, is the percentage of organisms of a particular genotype that will survive and reproduce, compared to others. The selection coefficient, therefore, is the comparative percentage that will not survive and reproduce (selection is considered as something that works against the organism). Usually, the coefficient of selection, written as s, is treated as something between 0 and 1. This means in effect that 100(1-s) percent are going to survive and reproduce. You can then factor this cause into the Hardy–Weinberg Law and show how selection will affect the percentages of a certain gene over one, two, or more generations (Figure 5.2).

The Level of the Gene

All of this is pretty elementary. There are a couple of questions that come straight up, however. First, what about selection considered as something that operates at the level of genes rather than of organisms? Darwin talked about organisms battling it out in the struggle for life and mates, and surely he was right to do so. A couple of stags fight, and then the winner copulates with the females. Of course, everyone has genes, but it is not the genes battling it out. Even if one agrees that the genes themselves have physical characteristics, and that these physical characteristics might themselves have effects, just like hands and eyes, that immediately affect survival and reproduction, at this point they are not being treated in this way. They are not being treated as physical objects, but rather as markers or tokens, carrying information. We know in fact that they are long molecules of DNA, but they could be made of cream cheese for all that population genetics cares, so long as cream cheese can carry information.

Is this acceptable? In fact, some biologists (Gould [2002], for one) have argued that it is not, and that there is something inherently confused about population genetics. If natural selection is about organisms, then Gould and friends would have us talk about organisms. This is not a general opinion, and for good reason (Dawkins 1976, 1982; Sterelny and Griffiths 1999). Already our discussion of simplicity has primed us for what is happening here. Scientific theories really are not just disinterested reflections of absolute reality, whatever that might be. They are abstractions from and interpretations of that reality, as scientists try to build pictures or (as they call them) models, which capture part of experience that is relevant (and ignore the rest of experience that is not relevant) and work toward understanding – explanation of what there is and prediction about what there will be. In the case of population genetics, scientists focus on one aspect (and one aspect only) of genes, namely, that they are (as noted earlier) signs signifying information. Scientists are saying that if you have these genes (or this combination of genes), then this will translate out as a certain kind of physical feature, or set of features.

Of course, this may be wrong, or you may think that biologists should spell out how the translation occurs – later we shall encounter biologists who think that one should and who are in fact trying to spell out the translation. But this is another matter. The point is that the genes are not being set up in opposition to organisms – either it is genes struggling, or it is organisms struggling. The genes are being used as recorders of the information that makes for organisms, and in particular of the information that makes for successful or unsuccessful (fit or less fit) organisms. So, in a way, we have two levels of selection – the genes (genotype) and the physical features (phenotype). But they are not rivals. Richard Dawkins (1982) has spoken of the genes as the "replicators" and the physical organisms as the "vehicles," and that is a good way of thinking of things.

The proof of the pudding is in the eating. Can the approach of population genetics throw light on the physical (including the biological) world? We will turn in a moment to questions of evidence, but note that the gene level is not the unique level of abstraction. There is no reason why one should not capture the action of selection without mentioning the genes. It is not that you think that the genes do not exist or are unimportant, but rather (perhaps because you are so ignorant about the genetics) that you want to stay just at the level of physical characteristics. One can do this, and in fact some of the most interesting work of recent years has focused on this level of analysis, looking at ways in which behaviors can lead to stable situations, or equilibrium (Maynard Smith 1982).

Such systems, involving what are known as "Evolutionarily Stable Strategies" (ESS's), are quite simple in theory. Suppose you have two forms in a population, let us call them Hawk and Dove, representing ferocity and timidity (though in real life, apparently, doves are quite ferocious birds). Suppose that when Hawks fight, they go all out for a win. But such fighting has its costs, at least it does when you fight with other Hawks – even winners tend to get beaten up some. Suppose that Doves are into

	Average payoff against Hawk	Average payoff against Dove	Overall average payoff
Hawk	1/3(−100)	2/3(100)	33.3 = 1/3(−100) + 2/3(100)
Dove	1/3 (0)	2/3 (50)	33.3 = 1/3(0) + 2/3(50)

FIGURE 5.3. Matrix of Hawks versus Doves in an evolutionary stable strategy. This assumes that you start with one-third Hawks and two-thirds Doves, that a victory is worth 100 points but a lost fight costs 200 points, and that nonviolent competitors split the spoils.

nonviolent encounters – no costs, but no gains when you meet a Hawk. Attaching simple numbers to these various strategies (the math is the same whether you pretend that all members of the population are either Hawks or Doves, or that all members of the population sometimes show Hawk and sometimes show Dove behavior), you can easily show that under certain ratios you have a stable situation. Equilibrium. The percentage of Hawks is not going to rise in the population, but neither is the percentage of Doves going to decline, or vice versa (Figure 5.3).

What you have here is a situation describing selection working at the phenotypic level (the level of physical characteristics or associated behavior) rather than the genotypic level (the level of the genes). Or let us be more precise about this. What you have is a way of conceptualizing the situation, so that you are talking only of the phenotypes and not speaking at all of the information encoding genes. You do not deny that they exist. You simply do not talk about them, and instead concentrate on the physical features and behavior. In the earlier models, you ignore the details of the physical features and behavior, and concentrate on the information at the genetic level. These are not contradictory approaches. They are simply different ways of going at the problem. In both situations, some ways of doing things are better than others. And this means that natural selection is at work. (Even in equilibrium, some ways are better than others, namely, those ways that do not displace equilibrium.)

Tautology?

Move on to one of the most common charges made against natural selection (Popper 1974; Johnson 1991). It comes up so often that if it proves to be mistaken – and it is – it is important to find out why it is so often leveled. This charge is that selection is a tautology, or a truism – that it is no true empirical claim about the real world, but merely a play on words that (like a stipulative definition) cannot be false. How can this be? Herbert Spencer proposed, and in later editions of the *Origin* Darwin accepted, an alternative name for natural selection: the survival of the fittest. Ask now: who are the fittest? The response comes that they are those that survive. Which means that the survival of the fittest collapses into the true but unhelpful statement that those that survive are those that survive.

Now clearly there must be something wrong here, for, whatever it may be, natural selection cannot be a tautology. It is too easy for it to be false – and tautologies can never be false. Let us suppose you have two kinds of organisms. For convenience, let us pretend that they are humans – though anything else would do as well – and let us call them Biologists and Philosophers. Biologists are sex-obsessed. All day long, that is the only thing that they think of, and they spend every last moment of their time scheming for success. Philosophers think only of the eternal verities and give sex nary a serious thought. Now, no doubt there will be some potential mates who, despite all, will prefer the Philosophers, simply because they are so noble and so self-evidently not interested in their own reproductive ends. But, on average, does anyone seriously think that Philosopher genes are going to spread more rapidly than or at the expense of Biologist genes? The Biologist genes are fitter than the Philosopher genes; they are going to spread and survive better than the Philosopher genes. This is an empirical fact rather than a tautology. Logically, it is possible that Philosophers always win hands down. Because of the Biologists' success in the real world, you can label the Biologist genes as "fitter" than the Philosopher genes, and in this sense there is no doubt but that the fitter are doing better than the less fit – there is a definition here – but nevertheless, we are talking about a real process in a real world. Biologists are getting on with doing what comes naturally (to them), and Philosophers are not. There is differential reproduction.

So what is going on here? At least part of the problem (if that is what it is) seems to stem from the fact that so simple a mechanism supposedly does so much. People cannot believe that mere differential reproduction can create the incredibly complex living world that we have around us. It is just not plausible that the hand and the eye – let alone the brain – are the end result of something like selection. At the very least, the feeling is, the cause should be as complex as the result, and that means that whatever it was that made the brain should be on a par with advanced quantum mechanics, totally opaque to those without a doctoral degree in the subject, and preferably rather more. So the sense is that selection must be simply redescribing what is going on, rather than providing genuine empirical insight. To which one can only say, look and learn. Example after example of selection at work will be given, to convince the skeptical. Stay with us on the empirical front, and we will see what we can do.

Part of the problem comes because some kind of inductive assumption about the uniformity of nature is involved with our thinking about selection. This would explain why people like the late Karl Popper, who despised induction and abhorred uniformity-of-nature assumptions, were always so suspicious of natural selection. The point is that one assumes that Philosopher behavior and Biologist behavior can be generalized. The sorts of behavior that characterize philosophers occur over and over, and have the same effects. Likewise with biologists. Suppose you have Theologians and Sociologists. Then you expect that Sociology genes will bring on the same behavior as Biology genes, with the same results, and that Theology genes will bring on the same behavior as Philosophy genes, likewise with the same results. Of course, one always has some kind of "all other things being equal" or ceteris paribus clause built into one's thinking, and this can cause trouble because things rarely are equal. It is well known, for instance, that whereas philosophers live a life of poverty, theologians live high on the hog, with adoring believers ever ready to cater to their slightest whims. This obviously could affect reproductive possibilities and actualities. But the underlying claims are empirical, however hard they are to uncover.

More than this: there are many obvious cases where selection has brought on uniformity – same challenges, same solutions. The classic case is that of marine animals. You need a certain-shaped body with certain appendages to achieve the maximum effects, swimming quickly and able to move up and down and sideways in the water. Not only has selection found this body, but it has produced it repeatedly in many different kinds of animal – fish, reptiles, and mammals. The problems were the same, and the answers were the same (Figure 5.4). "Adaptation is a real phenomenon. It is no accident that fish have fins, that seals and whales have flippers and flukes, that penguins have paddles and that even sea snakes have become laterally flattened. The problem of locomotion in an aquatic environment is a real problem that has been solved by many totally unrelated evolutionary lines in much the same way." Which leads to an obvious

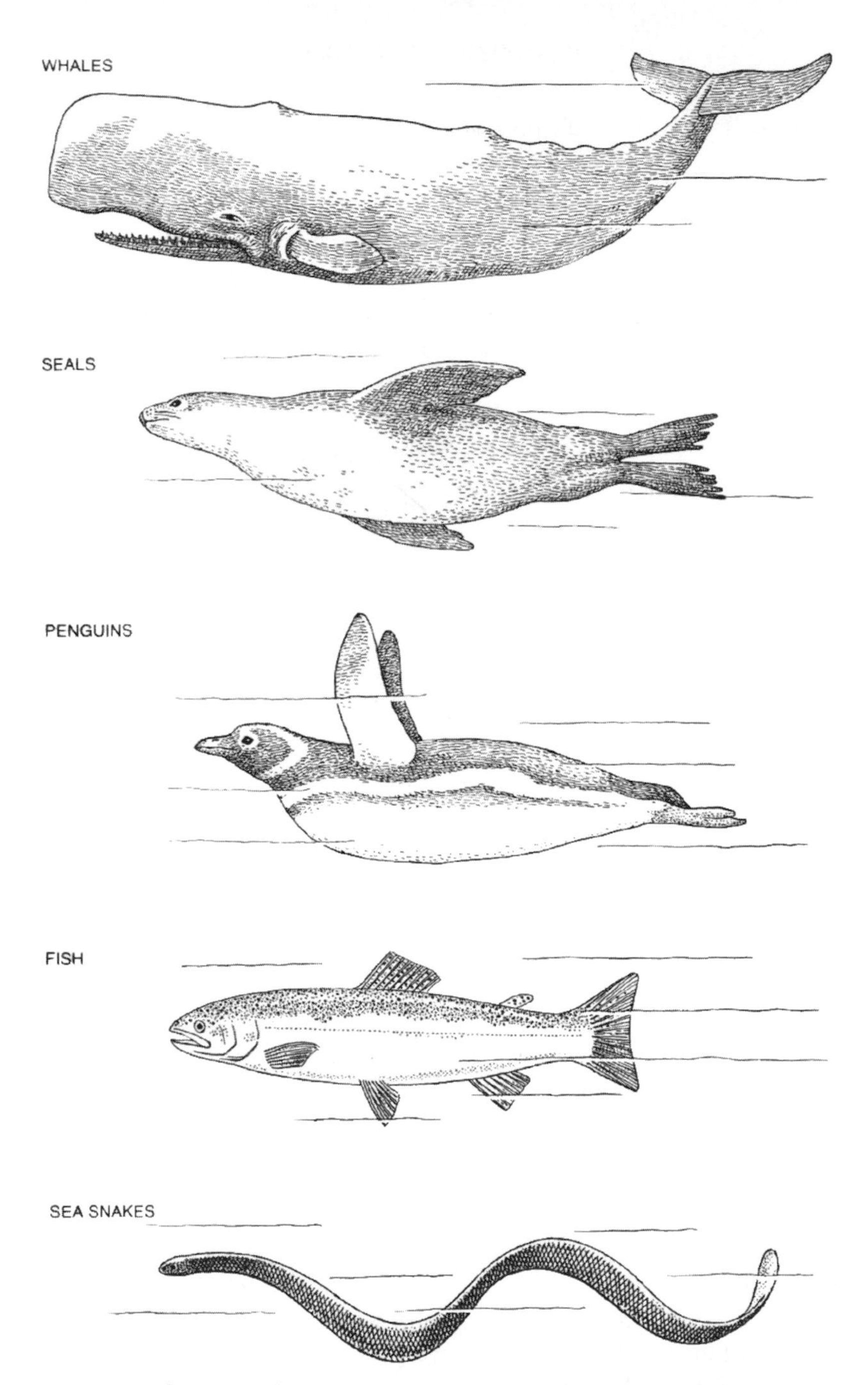

FIGURE 5.4. The same solution to the same problem as adopted by different organisms. From R. C. Lewontin, "Adaptation: the manifest fit," *Scientific American* 239(3) (1978): 213–31. Courtesy of Richard Lewontin.

conclusion: "Therefore it must be feasible to make adaptive arguments about swimming appendages. And this in turn means that in nature the ceteris paribus assumption must be workable" (Lewontin 1978, 228–9).

There is another aspect to the problem, one that is more a matter of the way of science than of natural selection in itself. Note that often, if not usually or always, scientists do not go to nature armed simply with one big overarching theory. Rather, as we saw in the last section, they build models of certain aspects of possible or potential experience (Giere 1988). Now, in an important sense, these models in themselves are not directly empirical – they are a priori. The models given earlier are cases in point. One decides to see how genes would behave over generations if one homozygote were less fit than the heterozygote or the other homozygote. One makes this assumption a priori. By definition, necessarily, the one homozygote is less fit than the other combinations. It is an analytic truth that the fitness of the homozygote aa is less than that of either heterozygote Aa or homozygote aa. Likewise it is an analytic truth that Hawks beat Doves, but that being Hawk-like has heavy costs when another Hawk is encountered. However, this does not make science in general, or natural selection in particular, simply a tautology – without empirical content. What scientists do now, having built their theoretical models, is to see if they apply in nature. (Often, of course, the building goes hand in hand with what is known or puzzling about nature.) This is where the empirical part comes in, as one makes claims – that might be true or might be false – about the applicability of the models to the real world (Beatty 1981; Thompson 1989). But true or false, that is another matter. Natural selection is more than a truism about the nature of fitness.

Empirical Evidence: Artificial Selection

The time has come to move from theory to evidence. Let us take another page out of Darwin's book. How did he make the case for natural selection? In the language of the Victorians, he wanted to show that it was a *vera causa*, a true cause, and to this end he employed two strategies. First, he tried to get at it by arguing that we have analogical evidence of natural selection in the breeder's world of plants and animals. Then he tried to provide a consilience, arguing that selection explains the facts of biology. Let us follow Darwin in his two strategies. (There was also the direct case for selection from the struggle for existence.)

Start with artificial selection. If people set about deliberately allowing some members of a population to breed, but restricting other members, do we get results? The answer is that of course we do, and in the chapter on Darwin, we discussed some of these labors – the effect of breeding on dogs, for instance. Turning to a much-discussed example from the days beyond the *Origin*, there was a selection experiment that went on for many (over sixty) years that tried to increase the oil content of corn (maize). The researchers (at the University of Illinois) started with 163 ears of corn, with an oil content in the kernels between 4 and 6 percent. They selected and bred from those with the highest oil content, and at the end of the experiment were growing corn plants with kernels having an oil content of around 16 percent – three times the amount that the best had when they started. Artificial selection had produced something distinctively new (Dudley 1977). As Darwin himself says, it is remarkable what can be done: it is almost as if breeders chalk an outline on the board and then go on and breed it.

Artificial selection by breeders exhausts Darwin's moves in the empiricist direction, but today we can go much further. We can look for experimental and natural direct evidence of selection in action. Taking things step by step, what about cases where the experimenter artificially provides conditions that might be expected to set up selective pressures? The experimenter is not actually choosing the organisms, but setting up a situation where certain organisms are expected to succeed and others to fail. The conditions might be entirely artificial and arbitrary, although more often the experimenter may be simulating conditions that are thought to have set up natural selective pressures in the past. Natural selection leads to adaptation, and hence what the experimenter is trying to achieve is adaptation in a population, brought on by those very factors that should kick-start the appropriate selective forces. Do these experiments in fact show change, as the Darwinian would expect? They most certainly do.

I have already mentioned Raphael Weldon (1898). He set up experiments simulating natural conditions and showed that selection brought on the changes he predicted. Today, one area on which there has been a huge amount of research and study is alcohol tolerance in fruitflies (McDonald et al. 1977; McDonald and Kreitman 1991; Ruse 2003). Normally, even slight concentrations of alcohol are fatal to fruitflies, but around breweries and wineries not only does one find flies that can tolerate alcohol, one finds flies that can exist only if there is alcohol. The Darwinian interprets this

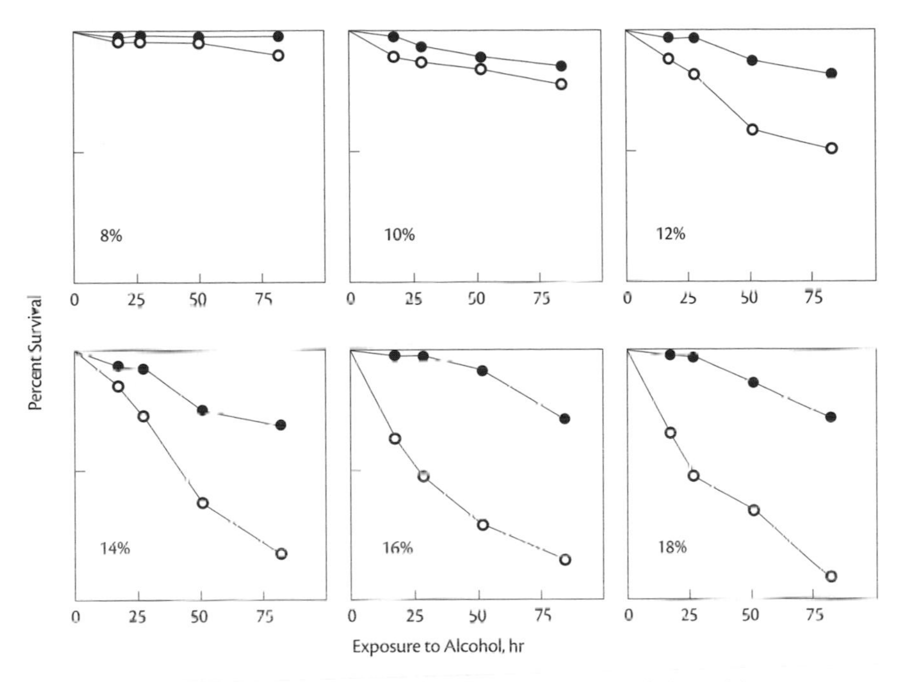

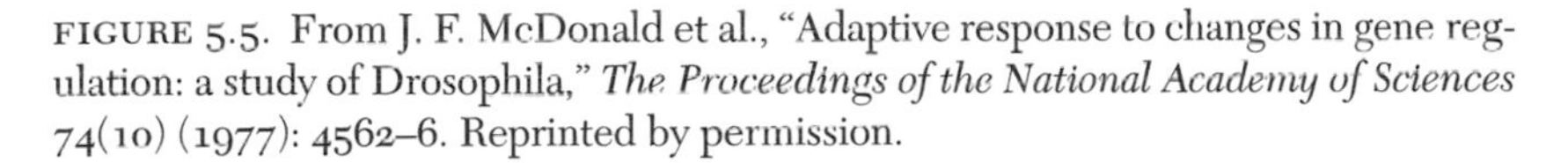
Survivorship of adult *D. melanogaster* flies from population selected for alcohol tolerance (●) and a population not selected (○) in six different concentrations of ethanol.

FIGURE 5.5. From J. F. McDonald et al., "Adaptive response to changes in gene regulation: a study of Drosophila," *The Proceedings of the National Academy of Sciences* 74(10) (1977): 4562–6. Reprinted by permission.

as a clear case of an adaptation brought on by selection, and experiments suggest that this is precisely so. John McDonald and his associates started with two identical populations of *Drosophila melanogaster*. One population was subjected, over twenty-eight generations, to intense (artificial) selection for alcohol tolerance. Those that survived alcohol environments were used to breed the next generation. The other population was not selected and used as a control. Then it was seen whether the selection indeed had an effect, and whether it produced what might be considered an adaptive ability to survive in and utilize alcohol-containing environments. The results were dramatic, especially as the alcohol levels were increased (Figure 5.5). Moreover, McDonald and associates were able to show that these differences were directly correlated to (and only to) an enzyme produced by what is called the Adh gene. In other words, they showed experimentally that the hypothesis that those flies able to utilize environments (like wineries) with high alcohol content could quite well have been selected to do so, and that the Adh gene could well have been a crucial causal factor.

Natural Selection

What about the action of natural selection in the wild? Or rather, what about direct evidence of the action of natural selection in the wild? When he wrote the *Origin,* Darwin did not have such evidence, but soon thereafter a close friend of Wallace, Henry Walter Bates (1862), produced a dazzling study showing that many butterflies mimic the members of other species. The reason is that those being mimicked are poisonous and hence unattractive to birds, whereas those mimicking are nonpoisonous and hence just pretending. Bates not only offered many examples of this kind of mimicry, showing that in some species of mimics it is more effective than in others, but also experimented, showing how birds will reject poisonous species and are fooled into rejecting nonpoisonous mimicking species. One is glad to be able to note that Darwin was very appreciative of this work, writing a (prudentially anonymous) enthusiastic review and finding Bates a permanent job.

Late in the nineteenth century, and right through the twentieth century, industrial melanism has been a star example of observed selection in action. In 1891, J. W. Tutt followed up on the observation of change with a Darwinian explanation.

> In our woods in the south the trunks are pale and the moth has a fair chance of escape, but put the peppered moth with its white ground colour on a black tree trunk and what would happen? It would . . . be very conspicuous and would fall prey to the first bird that spied it out. But some of these peppered moths have more black about them than others, and you can easily understand that the blacker they are the nearer they will be to the colour of the tree trunk, and the greater will become the difficulty of detecting them. So it really is; the paler ones the birds eat, the darker ones escape. But then if the parents are the darkest of their race, the children will tend to be like them, but inasmuch as the search by birds becomes keener, only the very blackest will be likely to escape. Year after year this has gone on, and selection has been carried to such an extent by nature that no real black and white peppered moths are found in these districts but only the black kind. This blackening we call melanism. (Tutt 1891)

FIGURE 5.6. Darwin's finches.

Tutt also made the (now confirmed) prediction that when Britain started to control pollution – forcing homes and factories to reduce the filth that they spew into the air – melanism would start to decline.

I will return to this example later. Here I want to look at the already-classic work done by Peter and Rosemary Grant on Darwin's finches, those little birds of the Galapagos that so excited Darwin himself when he visited the archipelago (Grant 1986, 1991; Grant and Grant 1989, 1995) (Figure 5.6). There were extensive studies of these organisms in the 1930s and 1940s, primarily by the English ornithologist David Lack, who, having first endorsed a non-selection-based position, then swung around and wrote a highly adaptationist account. But it is the Grants and their associates, starting in the 1970s, who have most carefully studied the birds and shown the action of selection in the wild. They have focused on a population of medium ground-finches, *Geospiza fortis*, on an islet (only a few hundred meters each way) known as Daphne Major. On average, there are about 1,200 specimens, and the Grants have caught and ringed them all. The birds can live for up to sixteen years, with a generation time of about four and a half years, so there is plenty of death and destruction going on. The question is whether this death and destruction is systematic, and if so, does it have selective effects?

The answer is strongly positive in both cases. The Grants have focused particularly on beak size and shape. First, are these features heritable? If there is no genetic causal connection, then selection could work away indefinitely without effect. But in fact, by measuring parents and offspring, it was seen that beak size and shape are strongly under the control of the genes. Big-beaked parents have big-beaked offspring, and so forth. What is the significance of beak size and shape? The birds eat nuts and fruits and the like. Big-beaked birds are going to be able to crack bigger and harder fruits and nuts than are small-beaked birds. Smaller-beaked birds, however, are going to be able to eat smaller seeds and the like. In other words, if there are mainly big and hard nuts and fruits, then the bigger-beaked birds are going to be at a selective advantage. If there is lots of everything, then probably the smaller-beaked birds are going to be at an advantage.

The Grants were able to test this, because there was a horrendous drought in 1977. Food supplies dried up, and the advantage shifted to those individuals who could exploit rarer or harder-to-access resources – namely big and hard nuts and fruits. Expectedly, the Grants found not only that the dead and emaciated birds tended significantly to be those with smaller and more refined beaks, but also that the average beak size shifted strongly over the next year to bigger and coarser beaks. There really was a gene-based shift, and it was in an adaptive direction that favored those birds able to access the more scarce resources. (There were no births in 1977, so this was something reflected in the already-living birds. However, when breeding recommenced in the following years, the young in turn reflected the selective pressures of 1977.) One should say that since all of this happened in the 1970s, there have been times of plenty, and that these in turn have favored smaller-beaked birds, which expectedly have made a comeback. But do not think that fluctuations of this kind should be taken to imply that nothing of significance happens over the long run. The birds rarely if ever return to exactly their original starting points. Rather, they pick up some new features and lose others. On average, over the past thirty years, the birds tend to be smaller and tend to have sharper beaks. One really does have evolution in action.

Speciation

A study like that of the Grants involves but a few years. In that time we do not see – do not expect to see – full-blown speciation or the evolution of brand-new adaptations and the like. But these are issues that cannot be ignored. Start with speciation, the formation of new groups of organisms that interbreed and that are reproductively isolated from all other organisms. There are two sides to Darwin's thinking on this issue. On the one hand, he saw variation across forms as being a result of the different opportunities that nature offers. There is sea, there is land, there is air – grab it! The pressure of the struggle is so great that organisms will be pushed to occupy or conquer new adaptive niches, and inasmuch as organisms themselves then create new niches – the tops of trees, the skin of mammals – so be it. This thinking was related to a metaphor that permeated Darwin's thinking – the division of labor. You need different and specialized skills to perform different and specialized tasks, and given that the world offers different and specialized opportunities, you get the evolution of forms with their own different and specialized adaptations. On the other hand, Darwin saw the actual breaking into different species as something essentially accidental. Groups are isolated, as with the finches on the Galapagos, and then when they come back in contact they have evolved so far apart that they cannot interbreed. Selection can harden or toughen the isolation, but it cannot really initiate it. It would never be in an organism's interests for hybrids to be sterile, but if hybrids are sterile, it would be in an organism's interests not to produce hybrids.

Turning now to the present, there are a number of questions (Coyne and Orr 2004). Can we produce new species artificially? Can we produce new species – at least reproductive isolation – artificially using selection? Can we devise experiments that simulate what we think happens in nature, that produce new species? Have we evidence of new species being produced in nature by selection? All of these questions can be answered positively. First, new species, by any method. It is well known that, in the plant world particularly, sometimes organisms are formed by combining complete sets of chromosomes from parents instead of the usual half – the process is called "autopolyploidy" when it happens within a species and "allopolyploidy" when parents are of different species (Stebbins 1950). The best-known case is the mating of a cabbage with a radish, first done by a Russian cytologist in the 1920s. The pity is that the hybrid has the root of a cabbage and the leaves of a radish. Second, new species thanks to selection. Two varieties (white and yellow) of corn (maize) were planted intermingled in a field. Each year, selection was applied, using those plants that showed lower rates of hybridization than their fellows. It took only five years for intercrossed matings in the white strain to decline from 35.8 percent to 4.9 percent, and in the yellow from 46.7 percent to 3.4 percent. There was no supposition that this would ever simulate a natural phenomenon, but it does show that if all you want is to get reproductive isolation, you can (Paterniani 1969). (Actually, in this case, it probably was akin to a natural phenomenon. There was no mystery to the decline in interbreeding. The white variety started to flower at an earlier time, and the yellow at a later. Thus, there was less chance of the fifth-generation populations producing hybrids. It is easy to think of natural conditions where climate and so forth might separate groups in this way.)

With these sorts of examples off our chest, turn now to more natural issues. It is believed that there are two basic forms of speciation – "allopatric," where the groups are physically (geographically) isolated, and "sympatric," where the groups overlap in some respect (Mayr 1942, 1963). Allopatric would be the kind of speciation supposedly bringing on different species in the Galapagos finches. It is thought that it can occur fairly rapidly in nature, but since "rapidly" means several hundred generations, it is not something we expect to see as a daily occurrence. If Darwin was right – and opinion is that he was – you are not going to see speciation occurring immediately for that end. Rather, it will be a by-product of selection for other factors. There are many reports of experiments where selection was performed for some end, working on different populations, and as a result the end groups showed some degree of reproductive isolation – usually behavioral rather than physiological, but isolation nevertheless. In fruitflies, for instance, selection for escape reaction, locomotor activity, temperature and humidity tolerance, adaptation to DDT (a very long experiment, involving around 600 generations), adaptation to acidity, carbohydrate seeking, photo- and geotaxis (attraction to light and gravity) all at the same time brought on degrees of reproductive isolation. Sometimes it happened rapidly in just a few generations; in other cases it took longer, as one might expect. The point is that it did happen. Expectedly, there are other similar experiments where such isolation did not occur, but given that it is a by-product and not itself a direct function of selection,

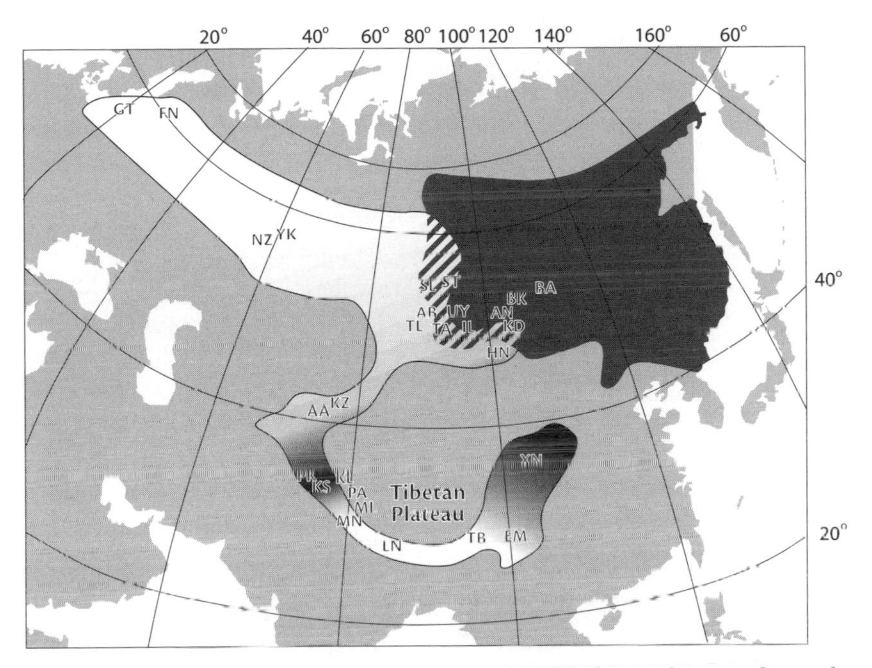

FIGURE 5.7. A ring of races in the greenish warbler complex. In the striped area, the populations overlap but do not interbreed. From D. E. Irwin, S. Bensch, and T. D. Price, "Speciation in a Ring," *Nature* 409 (2001): 333–7. Reprinted by permission.

this is just what one expects. (Coyne and Orr 2004 has a full discussion of these issues, with many references.)

What about nature? Even if you do not see it happening directly, you might expect to see it frozen in time, as it were. Do you see species in the making? The closest possibility would be "rings of races," where you have a succession of touching populations with interbreeding between members of such populations, but with the end populations touching and no interbreeding (Mayr 1942). (Strictly speaking, this is not quite allopatric, being more an intermediate form between allopatric and sympatric, known as "parapatric," where populations touch without overlapping) (Figure 5.7). There are many reported cases of such rings, but particularly today evolutionists tend to be critical about whether these can ever truly be shown to exhibit the supposed pattern. Probably some do, including the greenish warbler complex (*Phylloscopus trochilodes*) in Asia. It forms a ring of six subspecies (actually today broken because of deforestation), and molecular evidence shows that this is a ring where the neighbors are most closely related and where there is some interbreeding. Where the ends meet, the subspecies do not interbreed – they have different songs, and members of the opposing groups do not recognize each other (Irwin, Bensch, and Price 2001).

Sympatric speciation has always been more controversial, with some evolutionists (notably Ernst Mayr) denying that it can ever happen – such denial of course being the spur to other evolutionists (especially Mayr's students) to find it. Mention was made in an earlier chapter of a classic case in nature, the apple maggot fly that lives on hawthorn as well as apples. It should be noted that (inevitably!) some are not quite sure that this is as clear cut a case of sympatric speciation as it seems at first sight, but no one denies that in some wise we have species in the making or made. There have been many attempts artificially to simulate nature. One of the neatest had flies (*Drosophila melanogaster*) starting in a common place and then moving on and choosing habitats that distinguished between light and dark (phototaxis), up and down (geotaxis), and ethanol and acetaldehyde (chemotaxis). Flies were selected for breeding from those at the extremes (dark/up/acetaldehyde and light/down/ethanol) and quick breeders and slow breeders. The result was that rapidly (within thirty generations) one got flies choosing one habitat rather than another based on their history and not just randomly. Not every fly got every answer right – there were always flies from extreme parents who chose more moderate (that is, mixed) habitats – but the flies who did get things right came from the proper pedigree. In other words, there was mixing (necessary for sympatric speciation) and then assertive choice of habitat and mate (sufficient for sympatric speciation). (See Rice and Salt 1990.)

There is a huge amount more that can be said on the topic of speciation – we ourselves shall return to it in the next chapter. The point here is that there is reason to think that selection – perhaps sometimes combined with other factors to be discussed later – can do the job.

Adaptation

Turn next to adaptation, something that really does stick in people's craws. How can the eye have come about through a slow process like natural selection, with no guidance from above (or from anywhere else, for that matter)? At a psychological level – dealing with possibilities – taking a leaf from the *Naturphilosoph* handbook, Darwin had one effective answer. Even if you do not see the development of eyes through time, you do see the development of eyes through space. And these run in parallel. Today we have organisms that go from the simplest, most primitive of seeing organs – little more than slightly sensitive skin – to the eyes of humans and of hawks. You can put them in a row, and see that there is continuity with nary a gap, and that at each stage one has something that functions and proves its worth in the struggle for existence. This may not show that continuous change did happen in the past, but it does show that it could have happened (Dawkins 1986).

Proof that significant adaptation – large-scale adaptation – was caused by selection demands a switch to Darwin's other strategy, where we try to infer the action of selection by its results: consilience in the overall picture, "inference to the best explanation" or "abduction" in the particular instance. One has a phenomenon, and one asks what best explains it. In the case of adaptation, one is dealing with what Richard Dawkins has referred to as "organized complexity." Dawkins follows his fellow English evolutionist John Maynard Smith (1969) in thinking that this is "the same set of facts" that the religious "have used as evidence of a Creator" (Dawkins 1983, 404). One notes the existence of adaptation and asks how it can have come about. One eliminates the rival putative explanations, and one infers the working of natural selection. As Sherlock Holmes put it to his friend Dr. Watson, in Arthur Conan Doyle's "The Sign of the Four": "How often have I told you that when you have eliminated the impossible, whatever remains, *however improbable*, must be the truth."

A critic will quickly point out that one can never be truly sure that one has eliminated all of the impossibilities. Father Brown was a better guide here, for he knew that there is always some possibility of a rival explanation. Inference to the best explanation – abduction – is not a deductive argument. Given your premises, you could be wrong in your conclusion. But this is far from saying that such an argument is inadequate or time-wasting. Father Brown surely puzzled out the reason behind the strange behavior of Israel Gow. And a similarly strong case can be made in the selection–adaptation case. As Dawkins stresses – and here he is in a tradition that goes back to Plato and Aristotle – organized complexity has to have some explanation. It does not happen by chance (Ruse 2003). Take the eye. It is organized, it works. As Archdeacon Paley (1802) pointed out, you have to have an explanation.

> In crossing a heath suppose I pitched my foot against a stone, and were asked how the stone came to be there, I might possibly answer, that for any thing I knew to the contrary it had lain there for ever; nor would it, perhaps, be very easy to show the absurdity of this answer. But supposing I had found a watch upon the ground, and it should inquired how the watch happened to be in that place, I should hardly think of the answer which I had before given, that for any thing I knew the watch might have always been there. (p. 1).

A watch demands a watch maker. So likewise the eye – analogous to a human artifact like a telescope – demands an eye maker. Something so wonderful and complex does not happen by chance.

But what are the options? They have to be natural – according to law – or supernatural – meaning out of the course of nature. We have already opted for a naturalistic solution. What then are the options for a natural explanation? Selection or something else. But what something else? The obvious possibility that springs to mind is some kind of Lamarckism – the inheritance of acquired characteristics – and this is simply not true. Acquired characteristics are not inherited. The giraffe's neck does not get longer from ancestral stretching, and the blacksmith's arm does not get stronger from ancestral hours on the forge. In any case, as Darwin noted, Lamarckism cannot account for all adaptations. Granting that the sterility of hymenopteran workers is adaptive, it could not have been brought about by less and less reproduction. (Hymenoptera are the ants, bees, and wasps.)

Are there other options – those other possibilities to which Sherlock Holmes referred? Suppose there really is no direct cause – it is not so much that adaptation is uncaused, but that it is not caused by anything that is particularly directed. Mutations occur – or rather, major-effect-causing mutations ("macromutations") occur – and new adaptations spring into being. Now you are blind. Now (next generation) you can read *The Critique of Pure Reason* without the aid of spectacles. The trouble is – and this was the point made by Plato and Aristotle, long before Darwin – that random causes do not lead to organized complexity. In the language of modern science, major mistakes in gene copying do not lead to adaptation. Remember Murphy's Law – if something can go wrong, it will. As Dawkins argues, Boeing 747's crash and disintegrate. They do not lie in the wreckers' yard and suddenly reassemble when a tornado rushes through. This is not to say that one might not get some modified changes quickly. A 747 could be stretched significantly and still fly. A vertebrate might add on several vertebrae and still work. But a 747 does not spontaneously acquire wings, and neither does a bird or a bee.

There are those, and some are very vocal today, who think that there is more to be said on this matter. They think that in fact blind laws (beyond or without selection) can make for complex functioning. They argue that nature – physics and chemistry – has the power to self-organize. For the moment, I will leave this option undiscussed. (It will not be neglected or rushed by, I promise.) I will therefore assume that the (naturalistic) field is left to natural selection. If there is adaptation, then there was natural selection. But is there adaptation? Of course there is! The hand and the eye and wings and leaves and much, much else are adaptations. And it is precisely the ubiquity of adaptation that gives the Darwinian such confidence in the importance of natural selection. Inference to the best explanation. Precisely how widespread is the existence of adaptation is something to be discussed in the next chapter. Here, continuing the discussion, do note that the Darwinian does not simply assume adaptation and then explain it through natural selection, thus arguing for the importance of the Darwinian cause. This indeed would be assuming what you set out to prove. One has got to identify putative adaptations; one has got to show why they are adaptations; and one has got to relate them back as best one can to natural selection.

Sometimes you can do this fairly readily. Go back to the drunken fruitflies. John McDonald and his coworkers, remember, showed that flies with the ability to live in places with high alcohol contents could certainly have been selected to do so, and they also showed that the Adh gene could have been involved crucially as the cause. Independent studies have shown that it is precisely those flies with Adh genes that are found in alcohol-rich areas. In other words, the Adh gene gives rise to adaptive abilities to utilize alcohol as a foodstuff.

> Wineries have proved to be popular study sites for Drosophila researchers in many parts of the world. The interior of a winery cellar and the neighborhood of wine fermentation vats are good places to find high ethanol environments and to test hypotheses developed from laboratory data and from studies on natural populations in orchards. The distribution of Drosophila species inside and outside

wineries is in uniform accord with expectations; D. melanogaster [a species with high Adh activity] is found at much higher frequencies inside wineries than is the closely related but relatively alcohol-sensitive species D. simulans . . . " (Chambers 1988, 69).

We often do not have this much material to work from. But this does not mean that Darwinians can do nothing. There are at least two strategies that are important. Notice that an engineering metaphor seems appropriate for both. That is no surprise – nature had to put together those wonderful machines that we call organisms.

Optimality Models

Our hypothesis – our best explanation – is that natural selection has been at work. What would this mean? Let us take a best-case scenario and assume that selection has brought about perfect adaptation – it has "optimized" the situation – and that from here we can work out what is going on and why. Let us therefore build "optimality models" to explore cases of putative adaptation (Orzack and Sober 1994, 2001). The entomologists George F. Oster and Edward O. Wilson (1978) explicitly think of themselves as construction workers, as people making things that work. "In order to employ engineering optimization models the biologist tries to interpret living forms as in some sense the 'best.'" Of course, the trouble is with precisely what one means by "best" in a situation like this. "In effect the biologist 'plays God': he redesigns the biological system, including as many of the relevant quantities as possible and then checks to see if his own optimal design is close to that observed in nature." From then on, it is all rather a matter of trial and error, of putting the theoretical design model against the empirical findings. "If the two correspond, then nature can be regarded as reasonably well understood. If they fail to correspond to any degree (a frequent result), the biologist revises the model and tries again. Thus, optimization models are a method for organizing empirical evidence, making educated guesses as to how evolution might have proceeded, and suggesting avenues for further empirical research" (Oster and Wilson 1978, 294–5).

Is this an acceptable way of working? Some do not much like it at all. The population geneticist Richard Lewontin gripes that by "allowing the theorist to postulate various combinations of 'problems' to which manifest traits are optimal 'solutions', the adaptationist programme makes of adaptation a metaphysical postulate, not only incapable of refutation, but necessarily confirmed by every observation. This is the caricature that was immanent in Darwin's insight that evolution is the product of natural selection" (quoted in Maynard Smith 1978a; reprinted in Sober 1994, 99). Can things really be this bad? Let us look at Darwinians using optimality models trying to understand their subject.

Draw on an example involving foraging. Much attention has been paid to this problem. Animals have to eat, which means that they have to go out and find food – forage. You might think that it is all going to be rather simple. Make for the food and eat it, or carry it home and eat it later. But getting food can have costs – it may be a distance away or require energy to gather. Significantly, there may be danger in foraging – as you look for food, others also may be looking for food, and their gaze may turn to you. So one is going to have to start thinking in terms of optimal foraging strategies. How can you get the most food for the least effort or danger?

Suppose, to take a fairly simple example, an organism has two possible areas in which it can look for food – one area is rich in food; the other area has much less food. What is going to be the right strategy here? You might think that the animal will always go for the food-rich area. Suppose, however, that the food-rich area is in some respects much more dangerous than the food-poor area. Suppose that there is a major predator. How might things play out? One important variable might be growth. If the animal is small, better to stay in the food-poor area; but with growth, the animal will be less attractive to the predator, and hence a move to a food-rich area might make sense. You can complicate things a bit more. Suppose that if the animal is really small, then the food-poor area attracts other predators. Better to be in the food-rich area and hope to be too small to attract the attention of the major predator. Here, then, you might find with growth that the animal starts in the food-rich area, moves to the safer food-poor area, and then, as it approaches full size, moves back to the food-rich area. You can draw up a simple graph, and model the expected points at which changes of area might be expected to occur – what the optimum moves might be (Houston and McNamara 1999).

The evolutionary ecologists Earl E. Werner and Donald J. Hall found a way to test this hypothesis – a hypothesis that suggests that natural selection will have worked to optimize foraging behavior. They looked at the inhabitants of five small lakes in Michigan, studying the behaviors of bluegill sunfish (*Lepomis macrochiris*) and their densities and survivals with respect to their major predator, the largemouth bass (*Micropterus salmoides*). There are two main habitats, a littoral (beachy) one that has a fair amount of vegetation and a pelagic (deepwater) one rich in zooplankton. The bluegill spend the early parts of their lives in the deep water feeding on zooplankton. Then they move to the littoral areas and the vegetation. Finally, when they are older and above a certain size, they move back out to the centers of the lakes, to the deeper water and the zooplankton. All of this had to be established and tested. For instance, the bluefish were netted in both areas and measured and (for a select number) their stomach contents were observed and assessed, to see if those of a certain size really did seem to live and eat in the expected areas. The food availabilities of the different areas had to be checked, to see that there really was a differential there.

Also, it had to be established that the bass do not as easily eat bluefish in vegetated areas but can and do gobble them down in open areas, albeit with reduced success after the bluefish reach a certain size. This involved experimentation in artificial pond situations with vegetation (or not) and released fish, as well as observation in the wild. (One has the impression that a large number of students spent their summers in wetsuits below the surfaces of the lakes.) Other factors had to be considered and explored and sought – for instance, the significance of predator density (increasing density increasing the time spent in safe areas), and information on the genetic basis of the bluefish behavior. (One significant clue on this latter issue was the constant size across lakes of fish making the switch from pelagic to littoral habitats, where there was no likelihood that the switch was based on experience or information about predators.)

What were the overall conclusions from the study?

> The pattern of ontogenetic shifts in food and habitat use by the bluegill was quite consistent among the five lakes we examined. Upon leaving the nest, bluegill fry move into the pelagic zone to feed on zooplankton . . . and return to the littoral zone at ~12–14 mm SL [Standard Length]. After feeding in the vegetation for several years, bluegills again shift to feeding on plankton in the water columns, first above the vegetation in the littoral zone and then in the pelagic. . . .

> The data we present suggest that at least for larger bluegill a growth (or feeding) rate-predation risk tradeoff persists; during midsummer in our study lakes, the pelagic zone was energetically the most profitable habitat for all size classes of the bluegill, but the open-water habitat was also much more dangerous for small bluegills when bass were present. (Werner and Hall 1988, 1361)

Also, "this study illustrates the way in which the trade-off between growth rate and predation risk may influence the timing of ontogenetic habitat shifts in a species. Over a range of similar lakes, increases in bass densities delayed the shift by the bluegill to feeding on zooplankton: at the extremes, the bluegill spent 4 yr (Warner Lake) rather than 2 yr (Lawrence Lake) in the vegetation (i.e., over a 4.4-fold difference in mass)" (Werner and Hall 1988, 1364).

The researchers certainly felt that they had shown the worth of theoretical optimality models in exploring the reality of nature, and the effects of natural selection. Were they right in this, or is Lewontin's acid assessment better? Was it just an exercise in metaphysics, basically assuming what it set out to prove? The long and arduous work certainly does not come across as a "caricature," to use another of Lewontin's belittling terms. Let us admit that some kind of circularity was involved. The researchers assumed that selection would have made for an optimal use of time and effort, and built models on this assumption. They then tested these in nature. (In real life, one has a lot of work to adjust models as information flows in). Then, when the models did seem to map the reality of nature, they concluded that they had a situation confirming the workings of selection in nature.

Is this a vicious circle? Hardly! It is more the kind of feedback situation that is the norm in science – one puts forward an hypothesis, one tests it, and inasmuch as it works, one has renewed confidence in the hypothesis and the background against which it was formulated. Although in this case there was no explicit elimination of rival hypotheses, the kind of reasoning is very similar to inference to the best explanation – one is eliminating unstated rivals, such as the claim that everything is pure chance, or that different situations (lakes, in this case) require different analyses, or that everything rests on piscine culture and is learned, and so forth. In short, one has a paradigm case of using a theory – evolution through natural selection – to explore and understand nature, and inasmuch as it works (and in this case it does), one has feedback confirmation of the legitimacy of one's approach and background theory.

Reverse Engineering

Let us move now to situations where this kind of modeling is just not possible, where you cannot run the organisms through a cycle or two in order to test your hypothesis. What we have are cases of putative organized complexity and not much else – most obviously when we are working with the fossil record and trying to puzzle out the nature of features of long-extinct organisms. For instance, when we are trying to explain the parallelogram-shaped plates that run down the back of the dinosaur stegosaurus. The argument here depends on showing that one has a case of adaptation, and then arguing back to the existence of the one reasonable cause, namely, natural selection. But how does one show adaptation?

One shows adaptation by puzzling out function. There are at least two ways to do this. One is through phylogeny and homology. If one has two or three related extant (still living) groups, and they have features homologous with those in the extinct group, one can argue by analogy that the function of the features was the same as that for the extant groups. A case in point would be the feathers of Archaeopteryx. Why would one think them for flying? Because today's birds have feathers for flying. (Probably, thanks to other features, Archaeopteryx did not fly as well as today's birds. But it did fly.) The other way is by trying to build models analogous to the features of the extinct organisms, and arguing that the function of the models is shared by the extinct-organism features (Weishampel 1997). One "reverse engineers." So, for instance, if the complaint were made that the Archaeopteryx feathers could not possibly be used for flying – inadequate wings or whatever – then one might build models to show that they could indeed fly. Or possibly one might show that their feathers could do something else – heat the body or catch flies or whatever. Here inference to the best explanation may be very important, because there may be several rival hypotheses.

By way of example, let us look at the puzzling out of the function of the strange noses of the duckbilled dinosaurs (hadrosaurs) (Hopson 1975; Weishampel 1981). Flourishing some seventy-five million years ago, these were very peculiar-looking animals, with duck-like bills (very efficient for eating vegetation) and often very fancy crests on the skulls. In one group in particular (lambeosaurines), these crests were long, hollow growths, starting with the nose and going back across the head and sticking out at the back – stranger than something you see at Mardi Gras in New Orleans. Why did the lambeosaurines have these appendages; what was their function? Phylogenetic inference is not terribly helpful, because their nearest extant relatives – birds and crocodiles. – have nasal cavities but nothing like these. "The specific elaborations found in lambeosaurines have no homologues in birds or crocodilians. . . . In both living birds and crocodilians, the most important functions of the nasal cavity appear to be air conduction, conditioning, filtering, and olfaction. Given the repositioning and alteration of the nasal cavity in lambeosaurines, its functional significance is far from obvious from a solely phylogenetic perspective" (Weishampel 1997, 154).

Turning to reverse engineering, many hypotheses have been proposed for the crests: that they serve as snorkel tubes for the underwater-feeding reptiles, or as oxygen tanks needed for scuba diving; that they are sexual displays, like peacocks' tails; that they are for fighting, like deer antlers; that they cool the brain; that they are indeed used for smelling; that they hold salt supplies needed to regulate the blood contents of the brutes. Some of these hypotheses have not stood the test of time. It now seems that the duckbills did not spend massive amounts of time in the water looking for food. They were essentially land animals and had no need of snorkels. In any case, the cavities would hold far less air than would prove functional were the animals foraging beneath the surface. In some groups of hadrosaurs, the skulls are pretty much filled in, and they could serve as battering rams, or as weapons to be used in struggles of strength.

This leads one to think that there might be other aspects promoted by sexual selection in the hadrosaurs generally, and there are clues that this was so. Sexual display of the crests on heads is a good candidate – these are the sorts of baroque items one associates with this process – and some of the predictions seem well taken. It appears (to take three predictions) that these were social animals, an obvious corollary of sexual competition; that one almost certainly had sexual dimorphism, again something one expects given the struggle for mates (think of the differences between male and female elephant seals); and that the crests were most clearly defined and species-specific when there were other species nearby and hence a need for distinctive features.

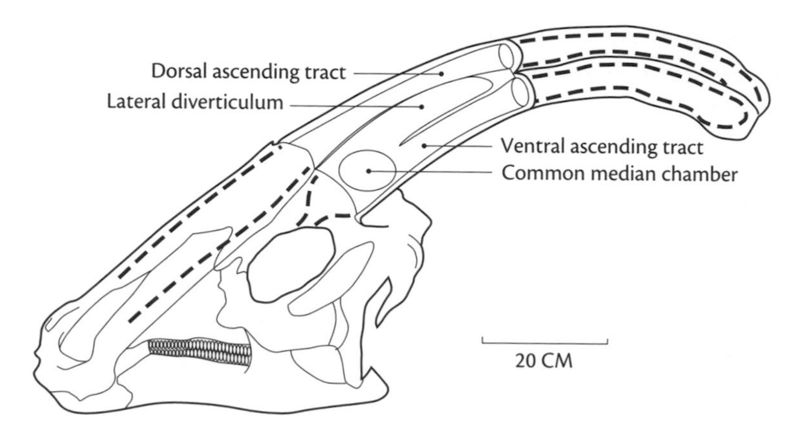

FIGURE 5.8. Cross section of hadrosaur (*Parasaurolophus walkeri*) head showing air flow passages. Adapted with permission from D. B. Weishampel, "Acoustic analyses of potential vocalization in Lambeosaurine dinosaurs (Reptilia: Ornithischia)," *Paleobiology* 7 (1981): 252–61. Courtesy of the Paleontological Society.

But this does not explain the internal structures, which do not seem to be directly linked to the various shapes on the outside of the skull. One hypothesis persists and has gained support. This is that these appendages were used for making sounds, allowing the dinosaurs to communicate with each other. Can this be so? Ask the obvious questions. First, would the appendages be able to make noise? The answer is that they almost certainly could. Unlike those of other animals, the air tubes within the appendages turn back on themselves, making for hugely long passages – we are talking a centimeter in humans and a couple of meters or more in these animals. We are talking trombone size and shape – as some early researchers noted, we are talking and looking like a medieval German wind instrument called the krumhorne. One is looking at something akin to the sound-producing apparatus of the trumpeter swan. One can work out the physics of the air flow through the nasal tubes, and it turns out that the dinosaurs could produce a huge amount of noise, particularly at low frequencies. (The trick is to get the sound resonating and hence much more bang for the buck.) Honking hadrosaurs, to coin a phrase (Figure 5.8)

The second obvious question is about hearing the noise. Fortunately, the fossil record is very good on hadrosaur ear bones, and here the phylogenetic approach does pay dividends. The bones are like those of birds and crocodiles, and since these latter are good at hearing, there is every reason to think that this was true of the dinosaurs also.

> In all, there appears to be good agreement among auditory anatomy, frequency transmission, and basilar membrane biomechanics for living archosaurs (birds and crocodilians). Furthermore, the anatomical similarities between the hearing apparatuses of these archosaurs and those of hadrosaurids strongly suggest that the latter had comparable hearing acuity. And this acuity compares well with the postulated vocal abilities of lambeosaurines. (Weishampel 1997, 157)

General opinion now, therefore, is that the lambeosaurines were walking college bands. Palaeontologists now understand some of the most peculiar and unfamiliar features to be found in the fossil record: a triumph of adaptationist thinking. More than this: a triumph of thinking about natural selection, and a major reason why Darwinians are so enthused about their mechanism. It is not just something sitting on a shelf, waiting to be polished and then worshipped. Like all of the very best mechanisms in science, it is a tool that can be used to further explanation and understanding. By thinking in terms of selection, and hence adaptation – design-like nature – palaeontologists could worry their way back to putative scenarios showing the function or purpose of hadrosaur crests, and the reason why the lambeosaurines had hollow passages up and down their skulls. This was not something just assumed, nor was it metaphysical, nor was it viciously circular reasoning. It was shown how the crests could function as noise makers, how this could work to communicate between animals, and why other explanations were less than adequate. An inference to the best explanation. Where once there was dark, now there is light. where once there was quiet, there is now sound! What more could one ask?

CHAPTER SIX

Limitations and Restrictions

> The homologies of process within morphogenetic fields provide some of the best evidence for evolution – just as skeletal and organ homologies did earlier. Thus, the evidence for evolution is better than ever. The role of natural selection in evolution, however, is seen to play less an important role. It is merely a filter for unsuccessful morphologies generated by development. Population genetics is destined to change if it is not to become as irrelevant to evolution as Newtonian mechanics is to contemporary physics.
>
> S. F. Gilbert, J. M. Opitz, and R. A. Raff (1996)

The three authors of this rather gloomy passage are among the leaders of a new subfield of evolutionary studies known as "evolutionary development" (or "evo-devo" for short). They represent the opinions of a sizable number of today's thinkers who believe that new developments in biological thinking represent a significant contraction of the scope and power of natural selection. It is the aim of this chapter to run through some of the claims made by these thinkers, as well as by others who – since the *Origin* – have tried to trim the sails of the Darwinian mechanism.

Is Natural Selection All-Powerful?

If natural selection is all-powerful, then we would expect to find adaptation everywhere, and that it always works perfectly. It was Stephen Jay Gould who launched the strongest attack on this picture of life, in a well-known article, coauthored by Richard Lewontin, "The Spandrels of San Marco." They argued strongly that evolutionists – Darwinians particularly – assume far too readily that living nature is adaptive, that it is functional.

They did not want to deny that the hand and the eye are adaptations, for clearly they are. But they felt that too often evolutionists slide into some kind of pan-adaptationism, of thinking that every last organic feature has to be functional, the product of natural selection. Referring to the Leibnizian philosopher in Voltaire's *Candide*, they accused evolutionists of Panglossianism, of thinking that these must be the best of all possible features in the best of all possible worlds. And to make the case complete, supposedly, evolutionists invent "just so" stories – thus named from Rudyard Kipling's fantasy stories – with natural selection scenarios leading to adaptation.

To counter this, Gould and Lewontin drew attention to the triangular decorative aspects of the tops of pillars in medieval churches (Figure 6.1). They argued that although such "spandrels" seem adaptive – areas for creative outpourings – in fact they are just by-products of the builders' methods of keeping the roof in place. "The design is so elaborate, harmonious, and purposeful that we are tempted to view it as the starting point of any analysis, as the cause in some sense of the surrounding architecture." This, however, is to get things precisely backwards. Referring to the spandrels in Saint Mark's cathedral, Venice, which carry mosaics of the apostles: "The system begins with an architectural constraint: the necessary four spandrels and their tapering triangular form. They provide a space in which the mosaicist worked; they set the quadripartite symmetry of the dome above" (Gould and Lewontin 1979, 148). Perhaps, argued the two men, we have a similar situation in the living world. Much that we think adaptive is merely a spandrel, and such things as constraints on development prevent anything like an optimally designed world. Perhaps things are much more random and haphazard – nonfunctional – than the Darwinian thinks possible.

Now, what is to be said by the Darwinian in response to this charge? Simply this: whoever doubted the point that Gould and Lewontin are making? It has always been recognized by evolutionists – certainly from the *Origin of Species* on – that however common or ubiquitous adaptation may be, it is only part of the story. The living world is not – cannot be – totally and completely adaptive. In fact, this is one of the strongest points against Paley's God. There is far too much wrong with the world – too many instances of malfunction – to think that a designer has been directly involved with making organisms. Perhaps Gould and Lewontin are right in thinking that, too often, adaptation has been assumed when it clearly

FIGURE 6.1. Spandrels in King's College, Cambridge.

does not exist. But the Darwinian would point out that it is often only by assuming adaptation that one can expect to bully through to the answer. What would have happened if paleontologists had given up before they started and simply assumed that the hadrosaur's nasal passages were nonfunctional? But, to return to the criticism, Gould and Lewontin are reinventing the wheel if they think that they are drawing attention to some new or hitherto-neglected problem or issue.

Nor, incidentally, is this to point to the inconsistency of Darwinians in their use of optimality models. If one builds such a model and applies it to nature, one is not thereby committed to the thesis that nature is always optimal. We have seen that modeling is a matter of trying to match nature, not assuming a priori that nature must match. If an optimality model works, then so far so good. If it does not, then one must build another model or look for reasons why it does not work. Is there some reason why adaptation is not as good as it might be? Does one find things that are nonadaptive, perhaps even positively harmful to their possessors? Sometimes, to take an example that Gould and Lewontin endorse, you just get things that do not fit or that lag behind in some way, and that selection fails to correct. The ornithologist Nicholas Davies studies European cuckoos and the birds that they parasitize (Davies and de L. Brooke 1988). He has shown that the cuckoos have many adaptations designed to make them successful in their aim of dropping their eggs into the nests of others – for instance, whereas most birds like to linger over the laying process, cuckoos are in and out in a flash. One area where there is extreme competition between cuckoos and hosts – what evolutionists call an "arms race" – focuses on egg color. Robins and most other birds are very good at spotting alien eggs, and cuckoos are very good at camouflaging their eggs to look like those of the hosts. The exception is the dunnocks, hedge sparrows. They are parasitized by cuckoos, but lack the ability to discriminate between their own eggs and those of others, and in turn the cuckoos make no effort to disguise their eggs. There is a clear adaptive breakdown here, for the dunnocks lose out.

The answer apparently is that dunnocks have been only recently parasitized and have not yet built up adaptations to the cuckoos. A clever experiment, using model eggs of various kinds, provided support for this belief.

> The cuckoo breeds from Western Europe to Japan but not in Iceland, where it is only a rare vagrant and has never been known to breed. Iceland does, however, have isolated populations of meadow pipits and white wagtails (of which the pied wagtail is a subspecies). We therefore took our model eggs to Iceland. The Icelandic populations bred at low densities and we had to work very hard to find nests but the results were exciting. Both the pipits and wagtails showed much less discrimination against eggs unlike their own than did members of the parasitized populations of these two species in Britain. (Davies 1992, 229–30)

Clearly here the power of selection has not proven all-powerful.

Sometimes we get adaptive failure – or at least less than adaptive excellence – because, as Darwin always stressed, one is not setting down a blueprint for excellence and then producing it. One is simply trying to do better than others (Ospovat 1981). As the old joke has it: it is not necessary to be super fast to escape the bear in the wood, merely faster than the chap next to you. There is no need of adaptive excellence in some absolute sense. Factors like this were surely important in the early years of the Cambrian explosion. It was not that the new complex forms were super-efficient. It was rather that they did better than others and had no more-complex rivals to wipe them out. Evolution is a bit like trying to cross a desert in an old car. If it breaks down, you have to find some way to get things going with what you have – rather like the astronauts in Apollo 13 – rather than going back to the garage or the drawing board and starting again. This means that life – like the old car – will reflect the contingencies of the situation. Thanks to selection, life may make moves that work but that in the fullness of time no one would think ideal or even the best. Take human birth. No one would say that this is a model of adaptive excellence. It is a matter of compromise. You have animals that are bipedal, and you have animals that put a premium on intelligence. You want the babies to have brains as big as possible by birth. And so what you get are difficult and dangerous births – not as difficult and dangerous as to rule out all successful birthing, but threatening to both mothers and babies nevertheless.

Sometimes we just get stuck with history or with the reflection of history. Take homologies, notably the vertebrate case highlighted by Darwin: the bones of the forelimb. The hand of a human, the leg of a horse, the flipper of a dolphin, the wing of a bird (and differently of a bat) are all built on the same pattern, the same archetype. There is no direct functional reason for any of this. It is part of history. But, complains the critic, is not this precisely the sort of factor that means Darwinism is exhausted? It is a version of this argument that motivates those quoted at the head of this chapter, although they want to put it in molecular terms. We have learned in previous chapters how deeply homology strikes at the molecular level. Physical homology links vertebrates. Molecular homology links animals as different as humans and fruitflies. Remember how the genes that regulate the sequential development of body parts – eyes, limbs, wings, bodies, and so forth – are virtually identical (with identical tasks) across the insects and the mammals (Carroll, Grenier, and Weatherbee 2001). To reaffirm these authors: you could not have greater evidence of shared origins if you had planned it yourself. But what of selection? If one admits in the case of skeletal homology that this shows constraint on selection and the production of the nonadaptive, how much more true this must be of molecular homology.

But does this render selection impotent or irrelevant? It does stress again the point made earlier, that selection does not start every new experiment from scratch, but cobbles together the working parts from what already exists. You have a set of genes that control development? Put them to work in fruitflies and humans! But this is a far cry from challenging the power or importance of selection. If the genes do the job for insects and mammals, why is this a mark against selection? Only if the genes do not do the job do you have problems. If it ain't broke, don't fix it. Let us suppose – although there is enough variation to suggest that this is not necessarily true – that the particular combination of genes that we have today is the only combination that would work for either insects or mammals. In this case, presumably, selection had to (and would have to) work with what is on offer – it could not be (would not be) all-powerful in the sense of being able to dictate one mechanism for flies and another for humans, and then deciding that the same mechanism is best in both cases. But this does not mean that selection has no power or that it does not do its job. Selection took what there was and molded it to different ends. The resulting adaptations could always be very good. No one denies that the human forelimb and the horse's leg work quite brilliantly. And selection deserves the credit. The same for the molecules.

Not that this is to deny that, more dramatically, sometimes nature just goes the wrong way, and we show a history that is not very helpful – adaptation that is really not as good as it might be. Consider the human male urogenital system, where, thanks to our evolutionary history, the sperm duct is rather like a garden hose that takes an unneeded loop around a distant tree, on its way from tap to nearby flower bed. The duct got itself hung over the urethra, and so instead of going directly from testes to penis, it meanders around the body before it gets on with the job (Figure 6.2). Hardly a triumph of plumbing, although note that there is nothing surprising about any of this, and certainly nothing that threatens Darwinism or adaptationism as such. No one is saying that the urogenital system is not adaptive or that you would be better off without it. The point simply is that you must think in a relativistic sense – is it better than the competitors? – rather than an absolutist sense – is it the best that could ever be?

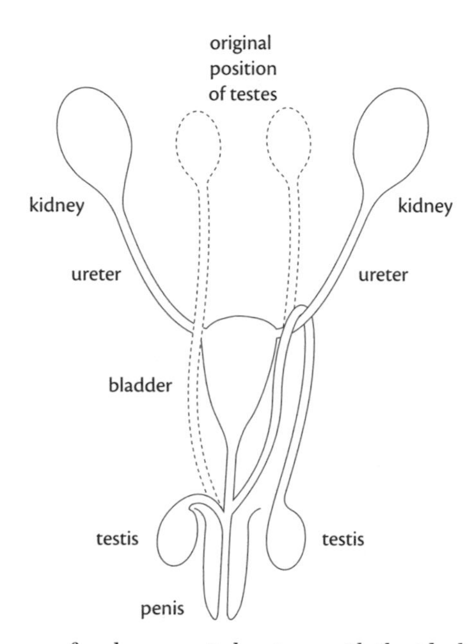

FIGURE 6.2. Diagram of male urogenital system, with the ideal on the left and the actual on the right. Figure 6.1 (p. 136) from George C. Williams, *Natural Selection: Domains, Levels, and Challenges,* copyright © 1992 by Oxford University Press, Inc. Used by permission of Oxford University Press, Inc.

The Maladaptive Costs of Selection

Thus far, we have looked at what might be thought of as failures of selection. Sometimes, however, selection itself can bring on a failure in adaptive excellence, especially when there are rival ends in view. It is well known that natural selection and Darwin's secondary mechanism of sexual selection can come into conflict. Selection in one sense may well favor one of these features, but selection in the other sense may not favor this feature at all. The peacock has magnificent tail feathers, thanks to sexual selection, but has trouble escaping from predators and hence is at a disadvantage from a natural-selection perspective. Another situation where selection and adaptation may get out of focus is where selection favors a feature at one stage in life but not at another. Allometric growth – where a feature grows at a rate different from the rest of the body – may be a key factor here. It is possible that the Irish Elk, notable for its too-magnificent antlers, fell victim for such a reason (Gould 1974). Selection favored rapid development and large antlers in adolescents, so that they

FIGURE 6.3. The overgrown antlers of the Irish Elk. From S. J. Gould, "Positive allometry of antlers in the Irish Elk," *Nature* 244 (1973): 375–6. Reprinted by permission.

could start breeding early and efficiently. Unfortunately, the animals went on growing and the antlers even more so, until they became positively disadvantageous (Figure 6.3). All-around perfection is simply impossible in such a situation. Sexual selection and natural selection were working to different ends.

Finally in this vein, let us mention that many genes have "pleiotropic" effects, making for or controlling more than one physical feature, and perhaps selection favors one feature so much that it overcomes the maladaptive nature of the other feature. Interestingly, there is one well-documented case of possible pleiotropy that seems involved in speciation (although this is not to say that speciation is necessarily maladaptive).

Plants are very sensitive to their surroundings, their soils particularly. One species of monkeyflower cannot normally grow on land that has a high copper content, but mutants are known (analogous to the wine-loving fruitflies) that grow well on nonferrous metal slag heaps. Hybrids between the normal plants and the mutants generally die, and it has been discovered that the same gene is involved in making for copper tolerance and for infertile hybrids (Macnair and Christie 1983). (One should add the qualification that it might not be exactly the same gene involved in both processes, but rather genes that are "linked," that is, close by on the same chromosome and generally transmitted together.)

Turn next to group situations, where natural selection can bring on horrendously maladaptive features. In the last chapter, we looked at how selection might bring on a balance or equilibrium – the case of the Doves and the Hawks. This was not unique or atypical. There are many, many ways in which selection might bring on balance between different members of a group. For instance, there could be selection for rareness. If a predator has to learn to recognize its prey, then (given two forms) if one is more common than the other, the predator will more often go for the former than for the latter. There will be selection in favor of the rare form until it starts to become more common, the predator changes behavior, selection switches the other way, and a balance ensues. Another famous way of selection bringing balance is if – comparing two alleles – the heterozygote is fitter than either homozygote. Intuitively, one can see that some copies of each allele will make it through to the next generation, however unfit the homozygotes may be. Even if one or the other (or both) homozygotes have no offspring at all, the heterozygotes will have offspring, and so copies of both alleles will be transmitted. In fact, mathematically it is easy to show that the same proportions will be transmitted in each generation. There will be balance or equilibrium.

The most studied case of balanced superior heterozygote fitness ("heterosis") occurs in humans, specifically, selection for malaria immunity (Cavalli-Sforza and Bodmer 1971). In certain parts of Africa, malaria is a great threat to human health, and people with one dose of a certain allele have a natural immunity to the disease. In other words, they are fitter than homozygotes without the gene. Unfortunately, having two doses of the gene, being homozygous for the allele, is fatal. Homozygotes suffer from sickle-cell anemia and (without dramatic medical intervention) die before they are four years old. In North America, this is a disease found

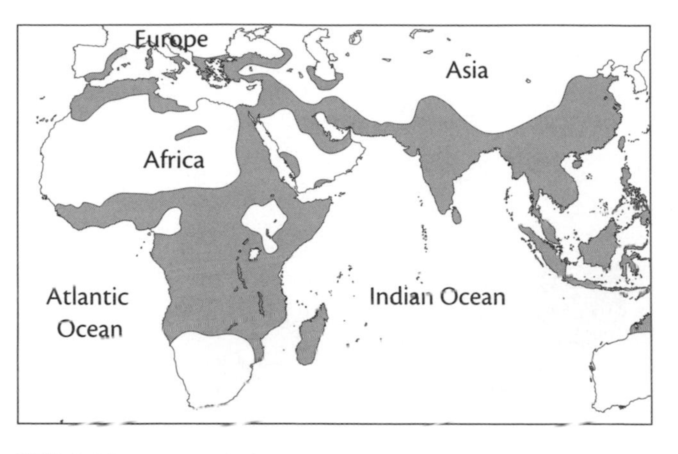

FIGURE 6.4a. Distribution of falciparum malaria. Courtesy of IBM Laboratories, San Jose.

predominantly among African Americans whose ancestors were brought from the afflicted parts of Africa. Now the disease lingers, for (in the absence of malaria in North America) there is no selective advantage to having just one dose of the gene (Figure 6.4).

Tempering the discussion at this point, one should say that it is a rare cloud, however dark, that has no silver lining. A consequence of the balancing effects of selection is that, at any time, you expect in the average population of organisms to find that there is significant variation. A major worry that people often have about selection is that they cannot see how it can be particularly creative. If mutation is relatively infrequent, small if it is to be effective, and random in the sense of not appearing to fulfill an adaptive need, how can major new adaptations ever come into being? It is all a bit like waiting for monkeys to type Shakespeare. But if there is always a lot of variation just waiting, then the possibilities are very different. A new predator arrives. Then, if there is not a gene for camouflage, there may be a gene for hiding; if not for hiding, then for fighting; and if not for fighting, then for getting the hell out of here as fast as you can. Think of an analogy. A student is asked to write an essay on dictatorship. If the only source material is from the Book of the Month Club, he or she could be very slow in getting the assignment done. But if the student can go to the library, and if the Hitler books are checked out, then there will

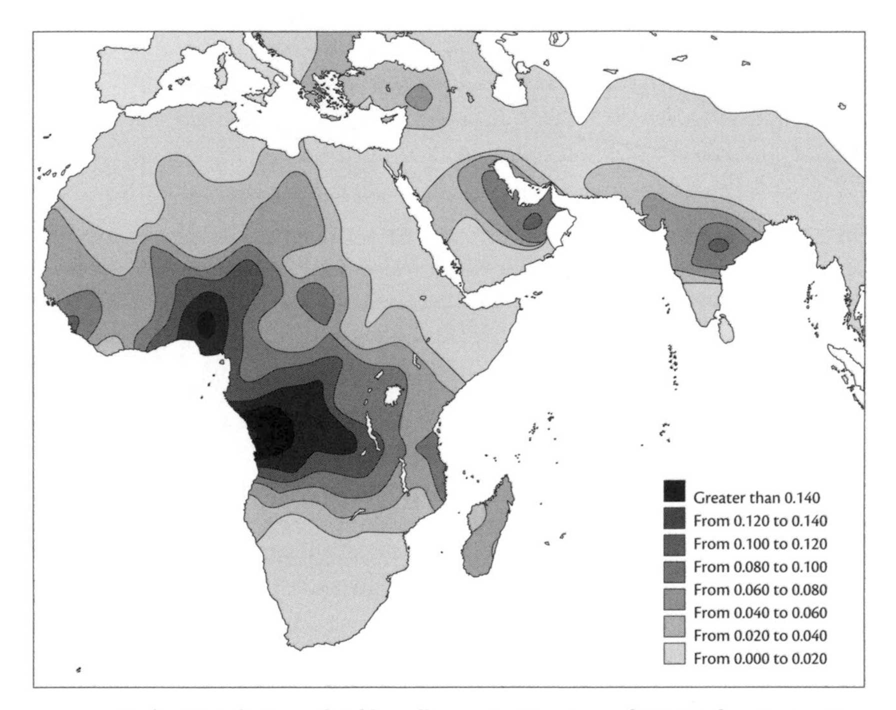

FIGURE 6.4b. Distribution of sickle-cell anemia. Courtesy of IBM Laboratories, San Jose.

be something on Stalin. And if not on Stalin, then on Mussolini, and so on. The essay can get written on time (Ruse 1982).

Levels of Selection

We saw in the last chapter that we can consider organisms at various levels of existence – at the level of genes (replicators) and at the level of whole organisms (vehicles). Is it possible that there are more levels above the individual and that selection sets one level against another? Ronald Fisher (1930) showed that sex ratios – generally 50:50 – are a result of an evolutionarily stable strategy. If a species has a preponderance of one sex over the other, then it is a good strategy for an organism to have offspring of the other sex. More opportunity, basically. If there are lots of girls around, then it is better to be a boy, and conversely. This would be the case until an equal balance was achieved and then equilibrium – stability – would ensue. But this balance is not necessarily something that is good for the group, or for all of the members of the group. At one level, females are better off without males because females have to do much of the child rearing. Indeed, apart from atypical groups like the birds, males usually do little or none. Hence, females reproducing without males – asexually – should be at an advantage over females reproducing with males – sexually. Asexual females are passing on all of their genes each time, whereas sexual females are passing on only half of their genes.

There has been (and still is) much discussion about sex and its putative evolutionary virtues (Maynard Smith 1978b; Ruse 1996b; West 2002). Most feel sure that sex does have some advantage, even for females. The English biologist William Hamilton suggested that it is a way to ward off parasites (Hamilton, Axelrod, and Tanese 1990). Sex keeps mixing up the genome, so that parasites – which typically evolve more rapidly than their hosts – are faced with a moving target and forever playing catch-up. There is a kind of arms race at work. But, even if this be true, one generally does not need as many males as one has. In a typical mammalian species, about 10 percent male is all that is required. This will keep genotypes shuffled quite nicely. Which raises the question of whether one can ever have a situation where selection works for the good of the group as a whole rather than just at the individual level? Intuitively, it seems that the answer should be yes. Apart from anything else, it at once gives us a reason why sex persists in groups despite the costs to individual females. A sexual group can spread advantageous new variations to all members of the group far more rapidly than can an asexual group.

Note that this question about individual versus group is not about natural selection or adaptation as such, but rather about the level at which selection operates. It is also not a level question of the kind we encountered in the last chapter in the case of gene versus individual. There we saw that selection can and does operate at both levels at the same time, without conflict. The levels are speaking to different issues. Here – as in the case of sex ratios – we have a potential clash between two levels, the individual and the group. Can selection favor group adaptations over individual adaptations (Brandon and Burian 1984; Sober 1984; Sterelny and Griffiths 1999)? The answer seems to be that no one wants to deny that "group selection" (to use the common name for selection making for group adaptations) can and probably does exist – especially in situations where group adaptations are really valuable and groups are small and transient, so that individual selection cannot really get a grip. But, generally, practicing evolutionists do not want to give group selection a very significant role (Reeve and Keller 1999). Generally, selection works for the benefit of the individual or his or her very close relatives. However desirable, group adaptations get corroded away or are unable to establish themselves because of the overwhelming effects of individual selection pressures.

Some complain that rejecting group selection comes more from a "reductionistic" philosophy favoring individuals and prejudice against a "holistic" philosophy favoring groups, rather than from reasoned consideration of scientific factors (Sober and Wilson 1997). We have already encountered the notion of reduction, and in a later chapter, we shall return to say more about reductionism and holism, and about their various meanings and plausibilities. Here let it be said simply that the charge of prejudice is not true. The sex-ratio case shows clearly how individual selection undermines group benefits. More than this, however, there are brilliant examples of individual-selection explanations of what prima facie seem to be paradigmatic effects of group selection. In other words, as a methodological strategy, favoring individual selection seems to pay dividends.

The classic example of such an individual-selection-based explanation is one that we have already promised to discuss: the sterility of the workers in the hymenoptera (ants, bees, wasps). What could be more obviously an effect of group selection than the fact that the female workers are sterile and spend their lives laboring for the good of the nest? Hamilton (1964a, b) pointed out, however, that the hymenoptera have an atypical mating system. They show "haplodiploidy." Females have both mothers and fathers – they are the result of the union of equal shares of DNA from female and male (and hence are diploid, with two half-sets of chromosomes) – whereas males have only mothers – they are born of unfertilized eggs, and hence have only the DNA of their mothers (and hence are haploid, having only one half-set of chromosomes). This means that, unlike normal reproductive systems where females are as closely related to daughters as to sisters (50 percent), in hymenoptera the females are more closely related to sisters than to daughters (75 percent; to 50 percent). Hence, from a selective viewpoint – where what counts is getting copies of one's DNA into the next generation – females maximize their own reproductive chances by raising fertile sisters rather than fertile daughters. A triumph

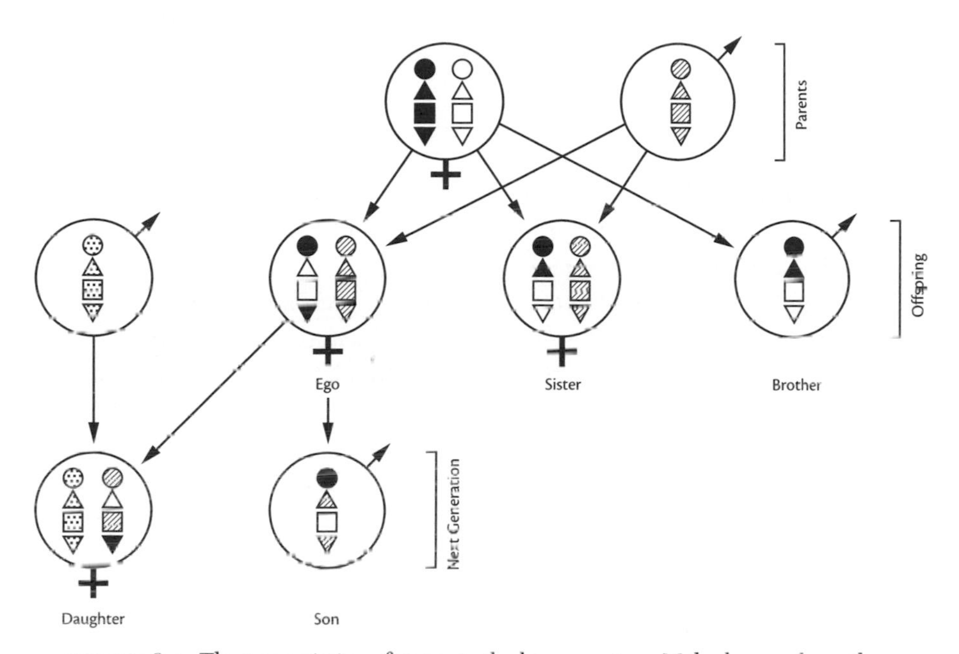

FIGURE 6.5. The transmission of genes in the hymenoptera. Males have only mother, females have both mother and father. In the offspring generation, the females are 75 percent related to each other, but only 50 percent related to their mother and to their daughters. From J. Maynard Smith, "The evolution of behavior," *Scientific American* 239 (1978): 176–92. Reprinted by permission.

of individual selection, because the infertility of the female workers is to their own adaptive advantage! (Figure 6.5)

In leaving this topic, one should add almost parenthetically that it is possible mathematically to show that processes like kin selection can be reconceptualized in terms of traits possessed by members of groups. However, it is a mistake to think of this in itself as revitalizing group selection, any more than it would be proper to think that because balanced heterozygote fitness occurs only in group situations, this denies individual selection, or that two organisms fighting each other manifests group selection. (Remarkably, this latter has been claimed, showing how far people will go to defend holism.) Whatever technical apparatus is used, it is most important historically to note that the early 1960s saw an absolute revolution in Darwinians' thinking about selection – a reversal, in fact, to much the way in which Darwin would have had us think – and that this has made a huge difference in the work done since then.

Drift

Selection – individual selection – is a major biological force. But are there systematic limitations on the power of selection? We have seen already that most Darwinians would agree that selection probably does not have major effects throughout the genome. It is this fact that makes possible the molecular clock. If there are no effects up at the physical (phenotypic) level, then selection cannot get a grip. For this reason, many gene ratios just "drift" aimlessly outside the control of selection. Although this is a theory that was conceived by the Japanese population geneticist, Moto Kimura (1983), he was building on an idea that goes back to the founding of population genetics around 1930.

There was a major difference between Fisher (1930) in Britain and Sewall Wright (1931, 1932) in America. Formally, even though they used very different techniques, Fisher and Wright agreed on the mathematics. This said, the two men's overall conceptions of the ways of evolutionary change could not have been more different. For Fisher, it was natural selection first, second, and third, as high as you want to count. Selection was always and uniquely the determining factor in evolutionary change. Mutations appear on a regular basis in large populations. Either they fail to make the grade, or they produce variations superior to those already existing and so rapidly become the norm throughout the group.

Wright's thinking was very different. For him, the key to change came from the fragmentation of populations into small groups – such fragmentation being a function of external causes. It is in these small groups that real innovative change takes place, and – here is the crucial part of Wright's theory – such change is in major part a result of random factors. This follows simply from the vagaries of breeding. One form, A, may be biologically superior (fitter) to another form, B. However, random mating in a small group, a process that will invariably produce some coincidences and distortions, might mean that B will prevail nevertheless – its genes will drift to fixity (to being the only representatives in the population). Then these new B-caused features, which themselves may not be directly adaptive, will join the general population as barriers break down and as subpopulations rejoin the main body of the group. In the full-population situation, the new features – produced by isolation and drift – may have advantages and thus may get selected and fixed through the whole group.

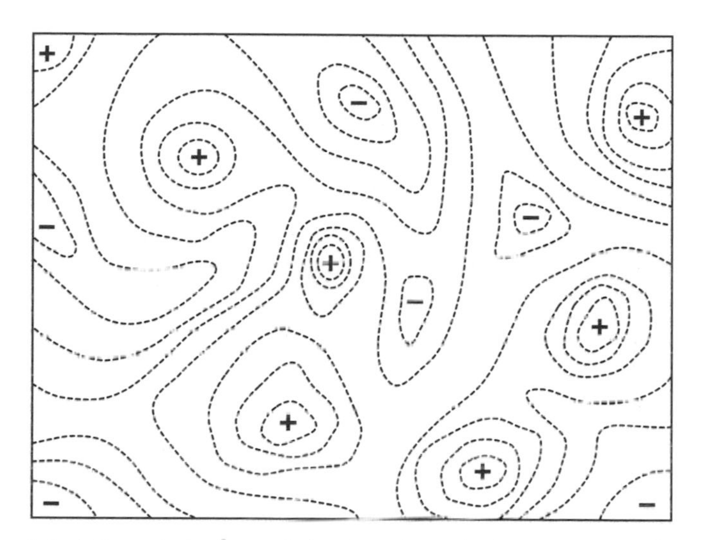

FIGURE 6.6. A "genetic landscape" showing the peaks (points of great fitness) and valleys (areas of little fitness). From Sewall Wright, "The roles of mutation, inbreeding, crossbreeding, and selection in evolution," *Composite Proceedings of the Sixth International Congress of Genetics*, 1932. Published by the Genetics Society of America.

Wright called his theory the "shifting balance theory" of evolution. What made his thinking really compelling, however, was not the mathematics (which at that time most people could not follow anyway) but a powerful metaphor he used to illustrate his theory. He invited us to consider organisms as if they were on an "adaptive landscape," with successes at the tops of peaks and failures down in the valleys. Evolution occurs when organisms move from one peak to another (Figure 6.6).

Let us consider the case of a large species which subdivided into many small local races, each breeding largely within itself but occasionally crossbreeding. The field of gene combinations occupied by each of these local races shifts continually in a nonadaptive fashion. . . . With many local races, each spreading over a considerable field and moving relatively rapidly in the more general field about the controlling peak, the chances are good that one at least will come under the influence of another peak. If a higher peak, this race will expand in numbers and by crossbreeding with the others will pull the whole species toward the new position. The average adaptiveness of the species thus advances under intergroup selection, an enormously more effective process than intragroup selection. The conclusion is that subdivision of a species into local races provides the most effective mechanism for trial and error in the field of gene combinations.

It need scarcely be pointed out that with such a mechanism complete isolation of a portion of a species should result relatively rapidly in specific differentiation and one that is not necessarily adaptive (Wright 1932, 68).

Wright did not want to deny that natural selection does operate and that there are adaptive features significant to organisms, but it is clear that he thought that many features simply have no connection to survival and reproduction. "That evolution involves nonadaptive differentiation to a large extent at the subspecies and even the species level is indicated by the kinds of differences by which such groups are actually distinguished by systematists." Apparently only when you start to get up to the subfamily or family level do you start to get adaptive difference. "The principal evolutionary mechanism in the origin of species must thus be an essentially nonadaptive one" (Wright 1932, 168–9).

Wright's theory is not very Darwinian. Natural selection does not play an overwhelming role. Genetic drift is a key player in Wright's world. However, although many of these ideas were taken up by later thinkers, especially by Theodosius Dobzhansky in the first edition of his influential *Genetics and the Origin of Species*, drift soon fell right out of fashion, thanks to discoveries that showed that many features formerly considered just random are in fact under the tight control of selection (Lewontin et al. 1981). Today no one would want to say that drift (at the physical level) is a major direct player, although, in America particularly, there has always been a lingering fondness for it. Many of the supposed supporting instances probably do not hold water – including Wright's own favorite, the desert plant *Linanthus parryae* (Coyne, Barton, and Turelli 1997). But there is some evidence for its existence. Caged populations of fruitflies have been shown – over generations – to vary in ways that one would predict from drift (Buri 1956). The same is true of some natural populations.

There is a significant corollary to drift, understood in the sense of a random determination of gene ratios rather than numbers fixed by selection. This is the so-called founder principle, something that could be a major player in speciation. Remember that the thinking is that much formation of new species comes about simply because groups get isolated and reproductive barriers accumulate accidentally as a result of the organisms in the groups evolving away from each other. Ask yourself how such isolation might occur in the first place. Quite probably, it is by just a small group of "founders" – perhaps even just one pregnant female – getting cut off from the parent population. Finches blown out to sea to the Galapagos, for instance. Ernst Mayr (1954, 1963) has pointed out that (because of the variation in all natural populations) such a small founding group of organisms will differ from the parent population. This entirely fortuitous or random difference – why these particular finches blown out to sea, rather than others? – will bring on a genetic revolution and subsequent physical change and speciation as the founders settle down into a stable group. Mayr always decried what he called the "bean bag" nature of population genetics – where the effects of genes are considered one by one, ignoring the real-life situation where genes interact one with another ("epistasis") and can have overall magnified (nonadditive) effects – and he used this as evidence of the significance of change that might be caused by the randomness of the founding population

The point is that although selection may take over immediately in the fashioning of the new group, it had no part in the original choice. As with drift itself, there is debate about how prevalent this phenomenon truly is. A very vocal group of population geneticists today argue that it is simply nonsense to say that their science ignores epistasis, and that although it may contribute to significant rapid change, it does not at all follow that it must. Of Mayr's notion of a genetic revolution, as of Sewall Wright's shifting balance theory, they say bluntly that "these arguments are no more convincing than claiming that pigs can fly, because parts of pigs (e.g., American footballs) have been seen in the air" (Barton and Turelli 2004). This said, there is some evidence of the founder effect, including the reduced variability of populations (species) of presumed recent vintage (Berry 1986; Caccone et al. 1996). Interestingly, this consequence may apply to humans. If, as many suppose, most humans stem from a population that came out of Africa, then you would expect more variation in the parent stay-at-home groups and less in the wanderers, and it is indeed true that there is more variation in African populations of humans than in those of other races (Tishkoff et al. 1996). More on these matters in the chapter on humans.

Constraints

Let us turn now to a recently much-discussed topic, namely that of constraint. Gould (2002) and other critics of the ubiquity of adaptation argue that often – too often – selection cannot work and adaptation fails because organisms are constrained in one way or another. "Organisms are capable of an enormous range of adaptive responses to environmental challenge. One factor influencing the pathway actually taken is the relative ease of achieving the available alternatives. By biasing the likelihood of entering onto one pathway rather than another, a developmental constraint can affect the evolutionary outcome even when it does not strictly preclude an alternative outcome" (Maynard Smith et al. 1985, 269). Thus defined, no one would ever deny that there are constraints and that they can be significant. In a way, our discussion of homology – skeletal and molecular – concedes this. Supposedly we do have historical constraints, because organisms' history – where they were located at a certain point in their history – constrains the subsequent moves and the power of selection. The vertebrate forelimb cannot be of any form it wants. It must be a modification of the proto-limb of the past. Likewise with the genes for development, if truly there was no other way in which things could be done.

What about other sorts of constraints? At a purely biological level, the developmental morphologist Rudolf Raff (1996) raises the issue of genomes. "Having a large genome has consequences outside of the properties of the genome per se. Larger genomes result in larger cells. Because cells containing large genomes replicate their DNA more slowly that cells with a lower DNA content, large genomes might constrain organismal growth rates. Cell size will also determine the cell surface-to-volume ratio, which can affect metabolic rates" (p. 304). Salamanders often have large genomes and thus are good organisms on which to test hypotheses about constraints. And there does seem to be some evidence of their operation.

> Roth and co-workers have observed that in both frogs and salamanders, larger genome size results in larger cells. In turn, larger cells result in a simplification of brain morphology. Thus, quite independently of the demands of function, internal features such as genome size can affect the morphology and organization of complex animals. Plethodontid salamanders share the basis vertebrate nervous system and brain, but they have very little space in their small skulls and spinal chords. (p. 305, referring to Roth et al. [1994])

Yet Raff has to admit that if there are constraints, they cannot be that overwhelmingly significant. The salamanders can still do some quite remarkable things. They certainly do not come across as organisms desperately functionally constrained. "These salamanders occupy a variety of caverniculous, aquatic, terrestrial, and arboreal habitats. They possess a full range of sense organs, and most remarkably, a spectacular insect-catching mechanism consisting of a projectile tongue that can reach out in ten milliseconds to half the animal's trunk length (snout to vent is the way herpetologists express it)." They have excellent depth perception, too. Paradoxically, in fact, their slow metabolic rate, brought on by large genome size, may even be of adaptive advantage. "Plethodontids are sluggish, and the low metabolic rates introduced by large cell volume may be advantageous to sit-patiently-and-wait hunters that can afford long fasts. Vision at a distance is reduced to two handbreadths, but since these animals are ambush hunters that strike at short range, that probably doesn't affect their efficiency much" (p. 306). All things considered, nothing here need keep the ardent selectionist from a good night's sleep. And if insomnia does threaten, then there is the additional bromide that apparently, if need be, the salamanders can start to bring down their genome size.

Turning from the biological to the physical, there are many constraints influencing organisms and setting limits on what organisms can and cannot do, and on where and when they should invest their energies. Take one of the most basic of all facts: size and consequent weight increase rapidly, according to the cube power of length or height. Consider two identically shaped mammals, and suppose one to be twice the height of the other. The taller mammal is going to be eight times as heavy as the shorter mammal. From a structural perspective, it has eight times the weight problem. Do not ask why you cannot build elephants as agile as cats. They are a physical impossibility. They would need far more support, and this would mean that they would need bigger and stockier (non-cat-like) bones (Vogel 1988) (Figure 6.7).

Another interesting calculation concerns what has irreverently been called the "Jesus number." What are the constraints on walking on water? A fairly simple formula governs the activity. Pushing up is the surface tension, γ, times the perimeter of the feet or area that is touching, l. Pushing down is gravity, which is a function of the mass, m, times the gravitational attraction, g, or restating in terms of density, ρ, times the volume, which is a function of l cubed. In other words: $Je = \gamma l / \rho l^3 g = \gamma / \rho l^2 g$. Since the surface tension, the density, and the gravitational constant remain the same, this means that the ability to walk on water is essentially a function of the perimeter squared. In other words, the smaller you are, the better off you are; and conversely, the bigger you are, the more likely you are to sink. This is no problem for insects,

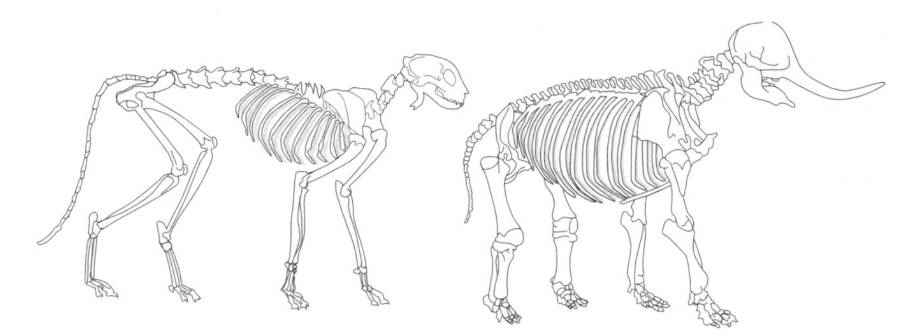

FIGURE 6.7. Comparison to cat (left) and elephant (right). If the cat were really as big as the elephant, it would need legs like the elephant's. From S. Vogel, *Life's Devices* (Princeton: Princeton University Press, 1988). Used with permission.

especially given that they have six legs and so have a long perimeter compared to the body size. Humans, however, are another matter.

> What would be the maximum weight of a human who could walk on water? My size-9 sandals have a perimeter of 0.62 meters each; that length times the surface tension of water gives 0.045 newtons of force, or 4.6 grams (less than half an ounce) of weight – 9.2 grams to stand (two feet in contact) or half that to walk. The theological implications are beyond the scope of the present book. (Vogel 1988, 100).

This book too!

Physical constraints are important, although there are times when it is not obvious that such constraints should really be called "constraints." John Maynard Smith and his coauthors have explored in depth the example of the coiling of shells in such organisms as mollusks and brachiopods. The coiling itself is fairly readily reduced to a simple logarithmic equation, and it is possible to draw a plane that maps the coiling as a function of the vital causal factors, particularly the rate of coiling and the size of the generating curve (Figure 6.8). Given such a map, one feature stands right out for comment: whereas for most shells the coils touch all the way from the center to the perimeter, some such shells coil without touching. There is a gap between the coils. Now, map the actual shells of a group of organisms, as in the particular case shown (Figure 6.9), the genera of extinct ammonoids (cephalopod mollusks). The isomorphism between the theoretical and the actual is outstanding.

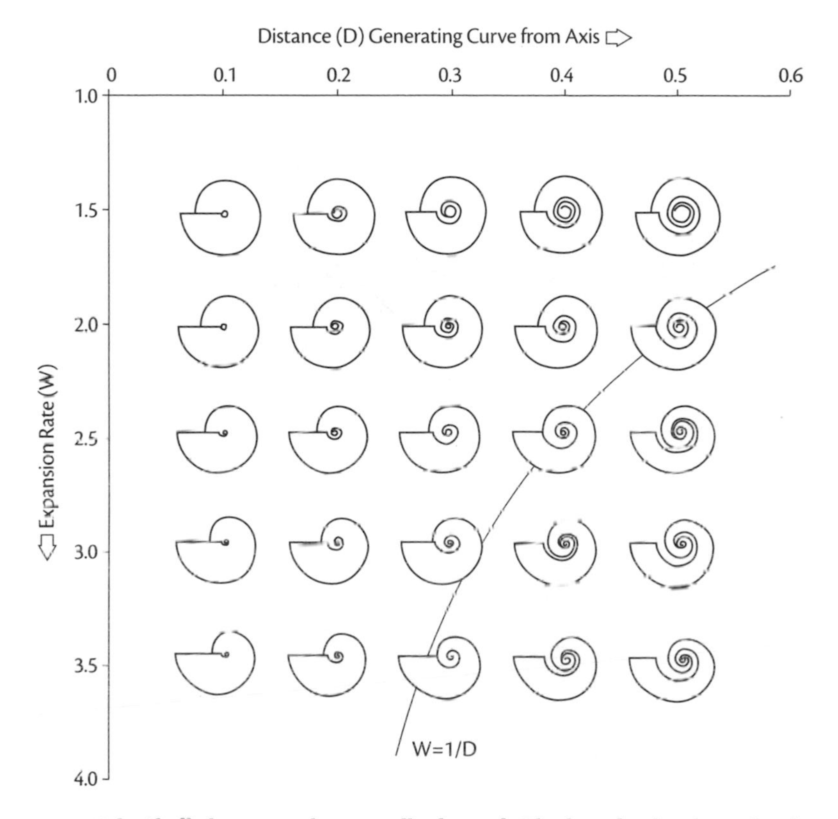

FIGURE 6.8. Shell shapes mathematically derived. The line divides those that have touching coils from those that do not. From D. Raup, "Geometric analysis of shell coiling: coiling in ammonoids." *Journal of Paleontology* 41 (1967): 43–65. Courtesy of the Paleontological Society.

As can be seen in [Figure 6.9], nearly all ammonoids fall on the left side of the curved line and thus display overlap between successive whorls. This is clearly a constraint in the evolution of the group but what kind of constraint? In this particular case, the answer is apparently straightforward (Raup, 1967). Evolving lineages can and occasionally do cross the line so there is no reason to believe that open coiling violates any strict genetic or developmental constraint. Rather, the reason for not crossing the line appears to be biomechanical. Other things being equal, an open coiled shell is much weaker than its involute counterpart. Also, open coiling requires more shell material because the animal cannot use the outer surface of the previous whorl as the inner surface of the new whorl. (Maynard Smith et al. 1985, 280)

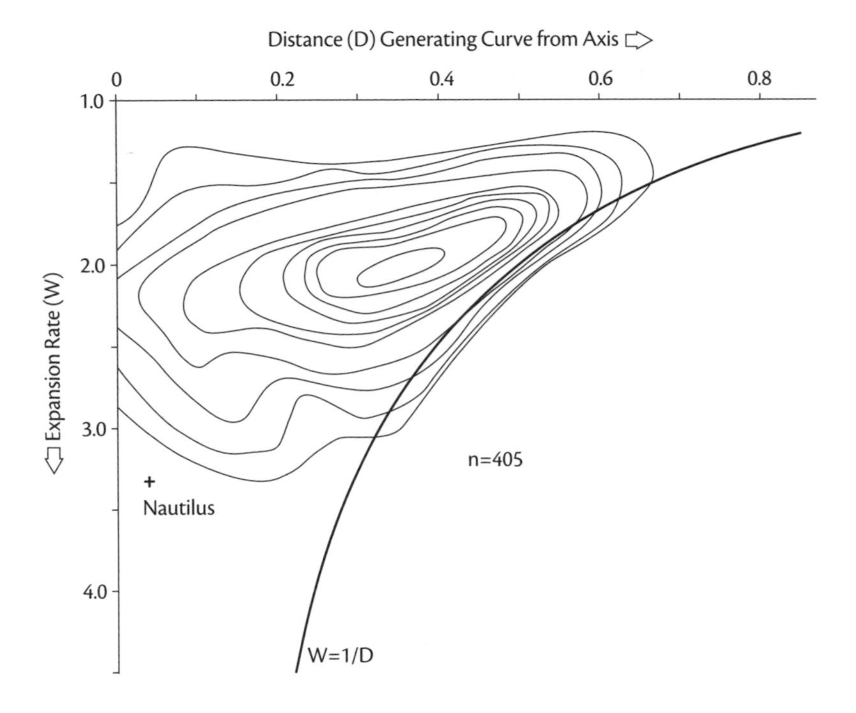

FIGURE 6.9. Real-life shells, mapping the mathematically derived shells. From D. Raup, "Geometric analysis of shell coiling: coiling in ammonoids." *Journal of Paleontology* 41 (1967): 43–65. Courtesy of the Paleontological Society.

Even the exceptions prove the point. The shell of the living pelagic cephalopod Spirula has a shell that coils but does not touch. Exceptionally, this organism carries the shell internally, using it for buoyancy. There is no need for strength. Maynard Smith and coauthors conclude "that the constraint against open coiling is an adaptive one brought about by simple directional selection." A conclusion that surely brings us full circle, for if constraints can be *adaptive*, brought on by selection, the whole argument against the ubiquity of adaptation is knocked sideways somewhat.

Punctuated Equilibrium

Move on now to another high-profile challenger to the role of natural selection. Notoriously, Gould – together with fellow paleontologist Niles Eldredge – was the author (in the early 1970s) of the theory of "punctuated equilibrium," a theory that argues that the history of life does not show the

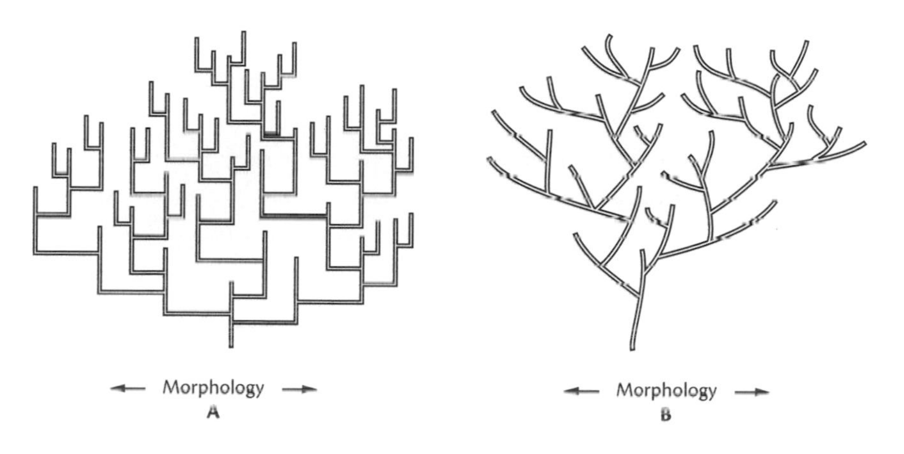

FIGURE 6.10. Comparison of jerky evolution (punctuated equilibrium) to smooth evolution (phyletic gradualism). Redrawn by Martin Young.

smooth gradual transitions that one expects from the action of unaided natural selection. The record is much more jerky, with rapid transitions from one form to another, and long periods of nonchange (stasis) between changes. Gould tied all of this in with his nonadaptive vision of the nature of life, and argued that when (as one should) we look at macroevolution rather than simply at microevolution, we see that there is a lot more to evolution than adaptation and selection (Figure 6.10).

Everyone agreed that Eldredge and Gould had put their finger on something – the question was, precisely what had they fingered and what were its implications? There was (and is) general agreement that they were right to point out that the fossil record does show long periods of nonchange, followed by rapid – instantaneous in the record – change from one form to another (Carroll 1997). Not always, admittedly: if you look at microorganisms you can often see changes occurring gradually over time. But if you look at mammals, for instance, generally what you see is one form and then another. The question was whether Darwinism could account for this. And the answer, in the opinion of Darwinians, was that it could (Charlesworth, Lande, and Slatkin 1982). Stasis does not necessarily imply no change at all – just no change recorded in the record – and if conditions do not change significantly, then stasis is certainly possible if not expected. Rapid change, as Eldredge and Gould themselves argued in their original exposition of their theory, could well be a consequence of the founder principle – admittedly not directly Darwinism, but involving Darwinian factors. A group moves to a new isolated habitat, and change takes place very quickly – too quickly to be spotted in the record.

Note how thinking like this meshes nicely with the assumptions of cladism, another movement just getting under way in the early 1970s. If the Eldredge–Gould picture is well taken, then most change is going to be associated with speciation, and given that most founder groups will probably be unsuccessful – too few founders, too harsh new conditions, and so forth – generally one is going to get bifurcation rather than trifurcation or more. Many finches may be blown out to sea, but it is unlikely that they will start many new species all at once. As it happens, many of the early cladists were not particularly Darwinian. Taxonomy is not an immediately causal science, and systematists have always regarded adaptations as more a hindrance than a help. One needs to dig beneath the surface to find true genealogical connections. Does this parallelism between cladism and punctuated equilibrium point the way to a divorce between the latter and Darwinism? Would it make more sense to free punctuated equilibrium entirely from natural selection and make room for outright anti-Darwinian factors? By about 1980, Gould was thinking that it does, and he himself started to favor some kind of "saltationary" theory, where change occurs instantaneously from one generation to the next (Ruse 1989). Adaptation was much downplayed, and the synthetic theory (as we saw at the head of the last chapter) was declared effectively dead (Gould 1980a).

This position was strongly criticized, particularly by microevolutionists (Stebbins and Ayala 1981). Gould was favoring the existence and significance of some kind of "hopeful monster" – a functioning organism that appeared spontaneously from parents very different. And this is simply without empirical warrant. No one wanted to deny that sometimes in development quite significant changes might take place in one or a very few generations – perhaps the lengthening of a neck or some such thing. But recreating an organism wholesale is just not the way that nature works. In the light of this onslaught, Gould pulled back somewhat (Gould 1982). Now he apparently did not want to suppose saltations, but rather that at upper levels there are other mechanisms that the microevolutionists miss. Which of course might be so, but until some convincing alternatives are supplied, Darwinians continue to argue that in important respects macroevolution is microevolution writ large. Natural selection working on random mutation is the key to evolutionary change, long term as well as short term. This is not to deny, as Gould stressed, that one cannot simply move from the short to the long, without taking account of much more than many had hitherto. In order to work on long-term patterns – the kinds discussed in Chapter 4 – one must have knowledge of large-scale phenomena, like extinctions and plate tectonics and much more. Then and only then can we start to grasp what has happened down through history. But the key point is that natural selection is a necessary friend in all of this, not a hated rival.

Order for Free

Perhaps Gould's fault was that he was too timid rather than too daring. Always he wanted to stay with biological causes, and this led to his downfall. As soon as he started speculating about things like saltations, conventional evolutionists were ready with their counterarguments. Perhaps the anti-Darwinian needs a much more radical attack on the need for natural selection or its place as the central mechanism of evolutionary change. Lamarckism will never replace natural selection, and neither will traditional, undirected variations. Yet have we eliminated the possibility that the regular, unguided laws of physics can do all that is needed – that they can entirely replace or significantly supplement selection? The focus now is on complexity – whether with Darwinians you think that this is adaptive complexity, or whether you think that it is just complexity and leave it at that.

Almost from the time of Darwin on – certainly through the last century – there have been those who think that natural selection is unneeded, a red herring (and a rather smelly one at that), and that the laws of physics and chemistry can do the job unaided. The leader of this school was a Scottish morphologist D'Arcy Wentworth Thompson. His work *On Growth and Form*, first published in 1917, is the founding bible. Thompson was one who took things to the extreme, denying that the organic world shows much design – he certainly had no time for the search for adaptation, thinking that this just holds biology back. "To seek not for ends but for antecedents is the way of the physicist, who finds 'causes' in what he has learned to recognize as fundamental properties, or inseparable concomitants, or unchanging laws, of matter and of energy. In Aristotle's parable, the house is there that men may live in it; but it is also there because the builders have laid one stone upon another" (Thompson 1948, 6).

Continuing: "Cell and tissue, shell and bone, leaf and flower, are so many portions of matter, and it is in obedience to the laws of physics that their particles have been moved, molded and conformed.... Their problems of form are in the first instance mathematical problems, their problems of growth are essentially physical problems, and the morphologist is, *ipso facto*, a student of physical science" (p. 10). Thus: "We want to see how, in some cases at least, the forms of living things, and of the parts of living things, can be explained by physical considerations, and to realize that in general no organic forms exist save such as are in conformity with physical and mathematical laws" (p. 15).

Thompson's approach to evolutionary questions was, to put it gently, somewhat fuzzy. Probably, despite his rhetoric, he was not entirely against adaptation as such. However, he surely wanted to downplay its existence and significance – at best, it was a corollary of the development of form, as governed by the principles of physics and chemistry. A paradigmatic case of nature's complexity that spurred Thompson to eloquence was the shape of the jellyfish. He saw this as a straight consequence of the physics of drops of liquid of one density falling in a liquid of a different, somewhat lower density. The patterns of the falling liquid are precisely the patterns of the organic, liquid-dwelling jellyfish (Figure 6.11).

> The living medusa has a geometrical symmetry as marked and regular as to suggest a physical or mechanical element in the little creature's growth and construction.... It is hard indeed to say how much or little all these analogies imply. But they indicate, at the very least, how certain simple organic forms might be naturally assumed by one fluid mass within another, when gravity, surface tension and fluid friction play their part, under balanced conditions of temperature, density and chemical composition. (pp. 396–8)

To be fair to Gould's credentials as an anti-Darwinian, some thirty years ago he did write an essay appreciative of Thompson's labors (Gould 1971). But it has been others who have done the spadework, notably the Canadian-born, English-dwelling morphologist Brian Goodwin (2001) and (in America) the medically trained theoretical biologist Stuart Kauffman (1993, 1995). The latter particularly, with the driven dedication of a prince seeking the Sleeping Beauty, has cut through the thickets

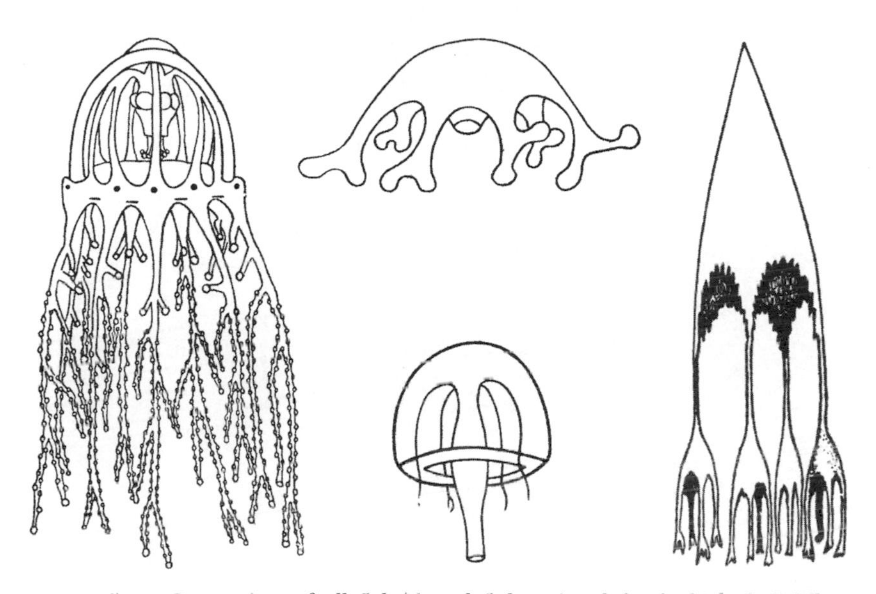

FIGURE 6.11. Comparison of jellyfish (three left forms) with fusel oil (alcohol) falling through paraffin (right form).

of "self organization," or (as Kauffman cleverly calls it) "order for free," finding physics lying there waiting for his kiss to wake it from its slumbers. "The tapestry of life is richer than we have imagined. It is a tapestry with threads of accidental gold, mined quixotically by the random whimsy of quantum events acting on bits of nucleotides and crafted by selection sifting. But the tapestry has an overall design, architecture, a woven cadence and rhythm that reflect underlying law – principles of self organization" (Kauffman 1995, 185).

Seizing on another example instanced by Thompson, today's formalists (as we may call them) invite us to consider the phenomenon of phyllotaxis, the pattern of clockwise and counterclockwise spirals shown by many plants, where identical elements are packed together (Figure 6.12). Sunflowers, pine cones, even the lowly cauliflower – they all exhibit this intricate pattern. There is no chance here. The pattern, phyllotaxis, is produced by the leaves or analogous plant parts appearing at the center (the "growing apex") and then, as it were, being pushed outward (Mitchison 1977). The appearing leaves or parts follow a twisted path (known as the "genetic spiral"), and if growth is constant, then the angle between successive leaves or parts is constant.

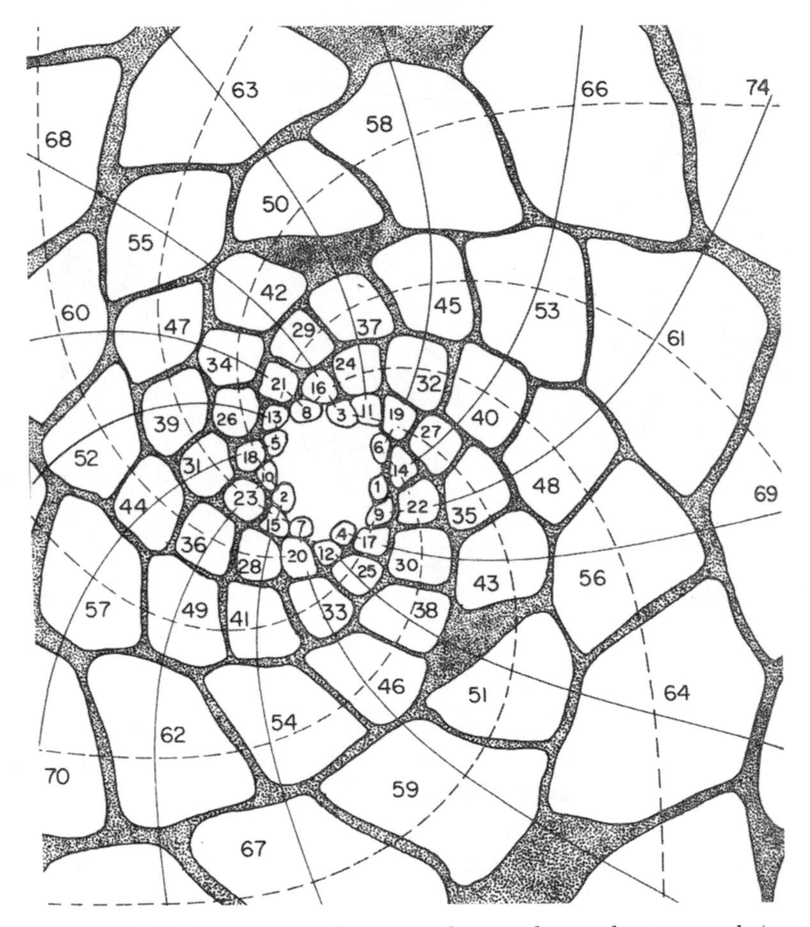

FIGURE 6.12. Leaf arrangements from a monkey puzzle tree showing spirals (parastichies). From A. H. Hall, *On the Relation of Phyllotaxis to Mechanical Laws* (London, 1904).

It is not the genetic spiral that catches the eye; rather, it is the criss-crossing diagonal spirals (known technically as "parastichies"). Pre-evolutionists realized that one can express phyllotaxis in mathematical form by means of a formula discovered by the thirteenth-century Italian mathematician Leonardo Fibonacci. Searching for a way to calculate the growth of the offspring of a pair of rabbits, he arrived at the series where any member is formed by adding together the previous two members of the series, starting with zero and one. Thus one has 0, 1, 2, 3, 5, 8, 13, . . . , or more generally, $n_j = n_{j-1} + n_{j-2}$. It turns out that, for any particular

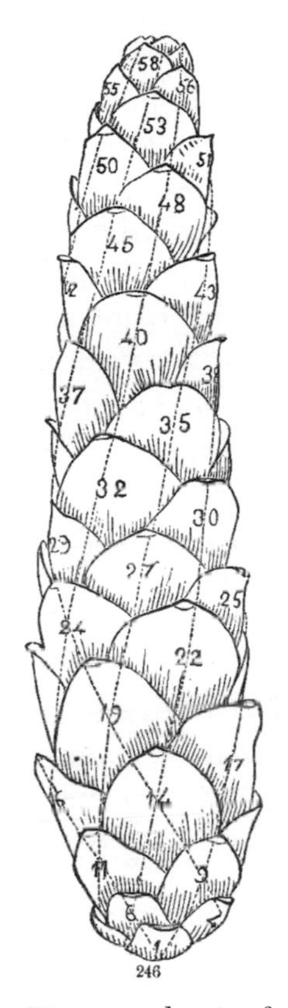

FIGURE 6.13. Pine cone drawing from Asa Gray.

species of plant, the numbers of parastichies, one set clockwise and one set counterclockwise, are always consecutive numbers in the Fibonacci series. In the stylized picture given in the diagram, the example is of an 8, 13 phyllotaxis. The way of calculating the measure is to use the order of production of the "contact" leaves on the same spiral. This is a measure based not on the order of production of the leaves but on the pattern itself, and refers to those leaves, along shared parastichies, that will be touching. Using examples furnished by Asa Gray in the sixth edition of his textbook *Structural Botany* (1881), the American larch produces a cone that is 2, 3; holly is 3, 5; and the cone of *Pinus strobus* is 5, 8 (Figure 6.13).

Darwinians have jumped on this phenomenon with joy. In the early 1870s, the American pragmatist Chauncey Wright argued that this kind of arrangement is the best way of exposing each leaf to the light, without undue overlap from its fellows. He argued that the differences between the various phyllotactic arrangements are so minute that they do not really matter that much. "To realize simply and purely the property of the most thorough distribution, the most complete exposure to flight and air around the stem, and the most ample elbow-room, or space for expansion in the bud, is to realize a property that exists separately only in abstraction, like a line without breadth" (Wright quoted in Gray 1881, 125). Formalists like D'Arcy Thompson and Brian Goodwin will have none of this. Thompson listed one objection after another. He thought that the differences between the arrangements is indeed significant; he sneered that the teleological intent is something that "cannot commend itself to a plain student of physical science"; he protested that there are all sorts of other ratios that could do the job just as well; he suggested that the plant could have taken alternative and superior routes in order to expose the leaves to sunlight, and so on and so forth. "We come then without more ado to the conclusion that while the Fibonacci series stares us in the face in the fir-cone, it does so for mathematical reasons; and its supposed usefulness, and the hypothesis of its introduction into plant-structure through natural selection, are matters which deserve no place in the plain study of botanical phenomena" (Thompson 1948, 953).

It will be no surprise to learn that these objections leave Darwinians unmoved. In their opinion, the formalists overlook the "obvious possibility" that "natural selection may universally favor close packing by phyllotaxis over alternative arrangements" (Reeve and Sherman 1993, 21). It is not to be denied that phyllotaxis is produced by development going down certain, fixed channels. How else would it occur? But form does not preclude function. Even if the patterns as such are fixed, nothing matters too much so long as other plant features can be varied in order to maximize leaf exposure efficiency. Which seems to be the case:

> Computer simulations indicate that phyllotaxy can influence the quantity of light intercepted by leaf surfaces. Model plants constructed with equal total leaf area and number differ significantly in flux, even when leaf-divergence angles are very similar. . . . Nonetheless, computer simulations indicate that a variety of morphological features can be varied, either individually or in concert, to compensate for the negative aspects of leaf crowding resulting from "inefficient" phyllotactic patterns. Internodal distance and the deflection ("tilt") angle of leaves can be adjusted in simulations with different phyllotactic patterns to achieve equivalent light-interception capacities. (Niklas 1988, 566).

The basic point is that however it is caused, complexity does not just happen. For its systematic appearance and persistence – as in the case of phyllotaxy – there has to be a tie-in with selection. It is not a question of form or function. Rather, it is function underlaid by form. Physics may produce the complexity. Natural selection cherished and refined the complexity, to its own needs.

Selection Slain?

What is our end point? It is just plain silly to say that Darwinism is an exhausted paradigm or that selection is a trivial cause of change – or even that it calls for significant revision or augmentation (Depew and Weber 1985, 1994). It is a powerful mechanism and has proven its worth time and again. It is not all-powerful. Natural selection has its limits – limits that have been recognized since the time of Darwin (he himself noted many of them) – but, taken as a whole, it is the key to understanding the organic world. There is no call for theory change yet, nor is there any prospect of such change in the near future.

CHAPTER SEVEN

Humans

I have to say that my chief feeling – I'll be honest about my chief feeling when I consider all this stuff – it's one of disdain. I don't know what to say, I mean, it's cheap!

U. Segerstrale (1986), quoting R. C. Lewontin

I suspect that truly there is only one reason why I am writing this book. If Darwinian evolutionary theory did not extend its grasp to cover us humans, no one would ever say anything nasty about it. Charles Darwin realized that humans would be the big issue and, in the *Origin*, for that reason – wanting first to put the basics of his theory on the table, as it were – avoided direct discussion of our species. Right at the end, not wanting to conceal his conviction that we are indeed part of the story, in the understatement of the nineteenth century, he wrote finally: "Much light will be thrown on the origin of man and his history." And with that, the floodgates were opened, and one and all – starting with "Darwin's bulldog," Thomas Henry Huxley, whose early book was called *Man's Place in Nature* – talked endlessly about what became known as the "monkey theory."

Darwinism and humans was controversial then. Darwinism and humans is controversial now. The opening quotation is from an interview given by the Harvard biologist Richard Lewontin, about the thinking on humans of his departmental colleague Edward O. Wilson. This is a mild comment compared to some of the things that are said. Lewontin and like critics do not have any doubts about the fact of human evolution. No one who accepts evolution denies that we humans are part and parcel of the picture. It is more a question of filling in the details. At the end of this book, I will turn to those who (for religious reasons) have trouble with the very fact. Here, keeping with scientific issues, I will look first at the path of human evolution. Then I will turn to Darwinian theorizing about the causes of human evolution and of our present nature as it has been shaped by evolution.

The Past

Even as the Victorians were jumping into discussion about human evolution and putative causes, the fossil record was starting to yield information about our past. The first examples of a human-like being, but (as they then saw it) more primitive, was being uncovered from the Neander valley in Germany. There was much discussion about the status of "Neanderthal man." Some thought him a new species. Others – Huxley was one – thought him somewhat different but not so much as to constitute a completely fresh line. After all, there were specimens living not that far from the heart of the Empire. "Ferocious gorilla like living specimens of Neanderthal man are found not infrequently on the west coast of Ireland, and are easily recognized by the great upper lip, bridgeless nose, beetling brow with low growing hair, and wild and savage aspect" (Grant 1916, 108).

The first true transitional fossil – the origin of the term "missing link" – was found at the end of the century, by the Dutch doctor Eugene Dubois (Shipman 2002). Other pieces of the puzzle fell into place in the first half of the twentieth century, and then – thanks particularly to the indefatigable Leakey family in West Africa – specimens of early proto-humans started to flood in. Let us lay out what we know in stark outline. The Age of Mammals began about sixty-five million years. The first primates go back at least fifty million years. As we come down to twenty or so million years ago, we are starting to get the first members of what we would now call the ape line. The big question, obviously, is precisely when the human line broke off from other animals, our closest living relatives. These, judging by anatomy, are the great apes: the chimpanzees (now divided into two species), the gorillas, and the orangutans. Fifty years ago, no one doubted that this break must have occurred at least fifteen million years ago, but then the molecular evidence started to come in, and it was found that we humans are incredibly close biologically to the apes. As noted in an earlier chapter, working with a version of the neutral theory, the molecular biologists put the date at only five million years ago or a bit more. Despite initial total disbelief on the part of conventional paleoanthropologists (students of the human record), this is now accepted by all as fairly definite. Equally fascinating is the question of our closest relatives – surely the gorillas and the chimps, with the orangutans a bit farther off. Most automatically assumed that the human line broke off and then the gorillas and chimps split, but now general opinion is that the humans and chimps went off alone and then split (Ruvolo 1994, 1995; Cann and Wilson 2003).

Probably the most famous fossil of them all is "Lucy," a very full skeleton of what is now classified as *Australopithecus afarensis*, discovered about thirty years ago in East Africa (Johanson and Edey 1981). (She comes from Ethiopia. There are other remains from Kenya.) About three and a half million years old, she and her fellow species members answered definitively one question that had fascinated students of human evolution from the earliest days, even before Darwin. The things that make humans really distinctive are the fact that we are bipedal – unlike the apes, we walk on our hind legs – and our large brains – 1,200 ccs or more, compared to the chimp at 400 ccs. Which came first? Lucy showed unequivocally that it was bipedalism. She (the famous specimen is thought to be female) was up on her legs, although detailed examination was to show that she probably did not walk as efficiently as modern humans and was probably much better at climbing trees than modern humans. Her brain, however, was chimpanzee size. (To repeat: this does not mean that it was a chimpanzee brain.) (Figure 7.1)

Recently, there have been some stunning finds that take us back from *Australopithecus afarensis*, possibly even to the point at which humans and chimps broke apart (Wong 2003a; Leakey and Walker 2003). *Sahelanthropus tchandensis*, so called because it was found in Chad (northerly Central Africa), dates from almost seven million years ago, and (only a skull has been found) has both ape-like features (brain case) and some more human-like features (teeth and shape of lower face) (Brunet et al. 2002). *Orrorin tugenensis*, from Kenya and coming in at six million years old, has a definite bipedal stance and way of walking, yet shows signs of ape-like upper features (the slant of the neck, particularly) (Pickford et al. 2002). *Ardipithecus ramidus kadabba* (the third term of this trinomial designates the subspecific classification) from Ethiopia (close by the home of Lucy) has toe bones intermediate between those of humans (upward tilt to joint surface) and apes (long and downward curving). It is dated at

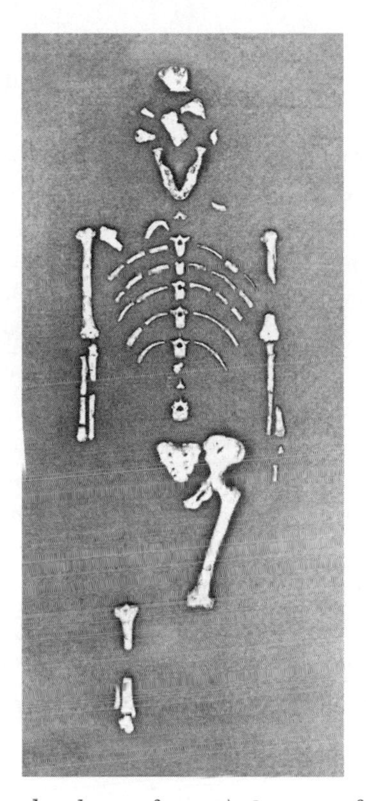

FIGURE 7.1. Lucy (*Australopithecus afarensis*). Courtesy of the Cleveland Museum of Natural History.

over five million years ago, with other specimens (*A. r. ramidus*) nearly a million years younger. How one fits all of these specimens together is a matter of some considerable debate. However, these things resolve themselves, we are getting a picture that fits nicely with the evolutionary scenario, from ape ancestors to humans, with chimps going off on their own way, around the time that the molecular biologists insist.

The picture after Lucy is now much fuller also (Hartwig 2002; Tattersall 2003a, b). Traditionally, one went from *Australopithecus afarensis* to the genus *Homo*, first to *H. habilis* (about a million and a half years ago), then to *H. erectus* (about a million years ago), and finally to *H. sapiens* (about a half-million years ago). This was marked by an increase in brain size, and the beginning of tool use (flaked stones) some two million or so years ago – with subsequent refinement as the years passed (Mithen

1996). This is still a good basic picture, but now it is much fuller. If Darwinian evolutionary theory is true, then although it is certainly possible that one gets a single species evolving steadily through time, it is more likely that one will have constant speciation and groups going different ways and subsequent extinctions of the less successful – probably less successful when competing with their own near cousins. A wide and ever-increasing range of finds strongly suggests much branching. At times – especially around two million years ago – there were three or four different species of our relatives around. Some of these are even put in a different genus (*Paranthropus*), although this is probably more a conceit of the scientists than a true reflection of relationships. One doubts that there would be such taxonomic proliferation were one working with fruitflies (Figure 7.2).

Finally, we come down to the near present. The most stunning find of all is the most recent, the little beings from the island of Flores, part of Indonesia (Brown et al. 2004; Morewood 2004). These creatures, *Homo floresiensis*, were about three feet tall, with brains of about 380 cc – the same size approximately as Lucy. There is debate about their origin. The official story is that they evolved from *Homo erectus* – and so the brain as well as the rest of the body shrank. More study is needed, however, and there are those who think that the most recent ancestor with humans may be several million years old. *H. floresiensis* existed twenty thousand years ago, and there is anecdotal evidence that they were even around five hundred years ago, when the Dutch first arrived at the island. Stunning, and yet in another way, not so stunning. We have just seen that branching is commonplace. That we humans were not always alone is no surprise – it is to be expected. That these creatures are smaller than their ancestors (if this is truly the case) is not even a surprise – dwarfing on islands is a common pattern, and apparently these creatures (who made tools and used fire) hunted elephants, which were also dwarfed. So scientists are excited but not upset. All pointers are that *Homo floresiensis* is going to fit very nicely into the Darwinian picture. That is not to say that it (actually the best specimen seems to be a she) raises no issues of importance to us. It does, and these will be aired and discussed in due time.

Let us end this survey with the Neanderthals and their relationship to us (Wong 2003b). There is a violent split in the paleoanthropological community (Ruse 2000). Some, supporters of the "multiregional hypothesis," think we humans are descended from them (Wolpoff and

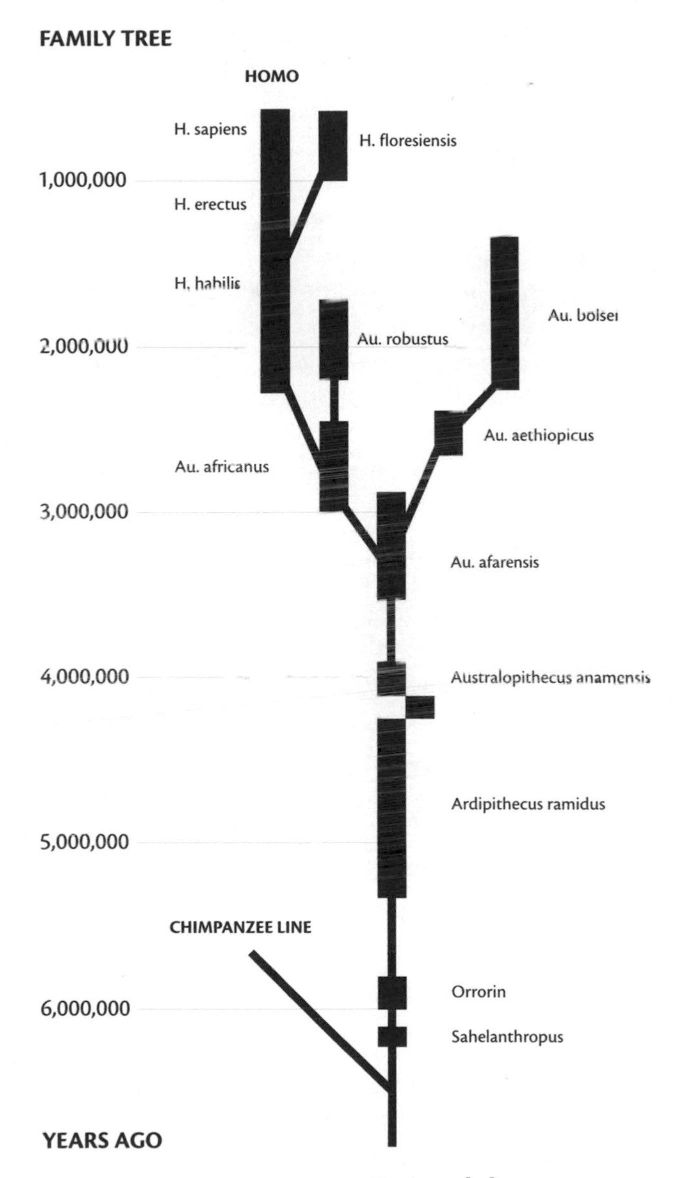

FIGURE 7.2. Human phylogeny.

Caspari 1997). Others, supporters of the "Out of Africa hypothesis," deny this strongly (Tattersall 2003b). No one disputes that we did come from Africa. Now, there is evidence of at least one shared ancestor no more than 130,000 years ago. This was a female, "Eve" (naturally!), whose mitochondria are shared by all of today's humans (Cann and Wilson 2003; Ayala 1995). (We are all descended from one female. This does not mean that we are descended from only one female, or that there is only one female from whom we are all descended, or that we are not all descended from one or a small group of males.) The big question is over the subsequent evolution. The multiregional hypothesis supposes that separate strands evolved in different parts of the world, in parallel. There was some gene interchange, but essentially the evolutions occurred apart. The European branch was the Neanderthal group, and this then evolved to become the modern human. We are not pure Neanderthals, but we do carry many Neanderthal genes. The Out of Africa hypothesis supposes that a group of humans came out of Africa and spread, wiping out all of the humans that existed in the various parts of the world. This included the Neanderthals in Europe, who are therefore now essentially extinct.

Probably, on average, Darwinians feel more comfortable with the Out of Africa hypothesis. This is not because they are committed to a violent picture of evolutionary change and this hypothesis predicts much conflict and extermination of preexisting groups, but rather because the idea of parallel evolutions ending at the same point makes them feel uneasy. It could happen, but it does not seem very likely. This does not deny absolutely that no modern human ever had sexual relations with a Neanderthal – there are some possible hybrid specimens – but rather that the Neanderthals were around and now they are not. Our ancestors drove them to extinction – either by bagging all of the resources or by fighting and killing them.

Causes

Why did human evolution happen? Why did we get up on our hind legs, and why did our brains get bigger? You might think that this was inevitable. Four legs good, two legs better. And surely a big brain is better than a small one. Does not nigh every science fiction scenario suggest that we humans are but a midway point, on the road to really big brains and computer-like intelligence? Well, right off, you can scotch all of that thinking. Lay aside *Homo floresiensis*. There is nothing wrong with being a chimpanzee – except the threat from humans. They are very well adapted for living in trees. We are not. They are safe from predators in a way barred to us, and when they want to move around on the ground, they can do so very quickly and efficiently by walking on their legs and the knuckles of their hands. Go to any zoo if you doubt this. And brains are even more problematic. They require a huge amount of fuel to keep them up and running. Moreover, all of that thinking is not necessarily such a good thing. Many mammals do pretty well without it. This applies especially to herbivores, which simply cannot get all of the high-quality food that big brains demand. In the memorable words of the paleontologist Jack Sepkoski: "I see intelligence as just one of a variety of adaptations among tetrapods for survival. Running fast in a herd while being as dumb as shit, I think, is a very good adaptation for survival" (Ruse 1996b, 486).

The initial moves to being human were triggered by climatic changes (Mithen 1996; Anton, Leonard, and Robertson 2002; Leonard 2003). The parts of Africa where our ancestors were living were more and more subject to periods of drought. This meant that jungles and forests started to dry up and to decline. The plains became, not just more available, but perhaps more necessary as places for living – if not all the time at first, then at least during the daytime. It could have been that our ancestors lost the battle with the chimp ancestors for prime jungle real estate. But if you are moving to the plains, then a number of reasons start to emerge for the move to bipedalism, and for other human-like features, including our relative hairlessness (Leonard and Robertson 1997; Stinson et al. 2000). Out on the plains you need adaptations for cooling – the loss of hair and (because of less exposure to the sun when standing) the possible move to the hind legs. Again, on the plains, being upright gives you much more ability to keep your eyes working to look around you and spot prey and predators. Remember, you are no longer hidden in the upper parts of leafy trees. The hands are obviously freed to be used to pick seeds and so forth. And perhaps of major importance is the fact that walking on your hind legs requires a lot less energy than running around using your knuckles for support. As you come out on the plains, you are moving from living much of your life in the trees to living much of your life on the ground. You need the ability to cover larger distances on the ground, and anything that reduced the energy required would have a strong selective advantage.

What about the brain? The key here is meat. It is this foodstuff that gives the kind of energy needed to support large brains. (For this reason, carnivores have on average bigger brains than herbivores.) The great apes are not strict vegetarians. They live mainly on fruits and shoots and other vegetable matter, but if they can get meat – a captured monkey, for instance – they eat it with relish. We humans have switched over to a very much more concentrated meat diet. Today, some groups (like the Inuit) eat meat almost exclusively, but all (including African hunter gatherers) eat much more meat than do the apes. Probably there was a cause-and-effect reaction at work here. As meat eating increased, the brains could get bigger and more efficient, and so the ability to get meat improved, and so the diet got richer and more energy-providing (Falk 2004).

Why was there the move to meat in the first place? Because it was there! As the jungles turned to plains, with their grassy expanses, the large herbivores moved in. Calories on the hoof, as one might say. It seems improbable that humans took to immediate and massive meat eating. It is not that easy to slice off a nice steak of elephant meat from a living beast. Most probably, we started as scavengers – primate jackals – looking for dead or dying animals, and following around and grabbing the spoils and remains from the killings of major predators like lions (Isaac 1983). Then, with the rise of brain power, proto-humans became more and more efficient at working together and at hunting in teams – grabbing the spoils before they were eaten or even catching or trapping the prey for themselves. At the same time, although there was some adaptive diversification among these creatures, thus reducing competition, there would have been competition among the different groups, and selective pressure pushing toward greater intelligence and ability to compete and conquer.

We do not yet know everything about human evolution. But the paths are being filled out nicely. Fossil finds are not ending. Indeed, they are increasing and improving. Likewise with causes. New techniques are coming on board – for instance, ways of determining diet from dental records and from bone compositions. Admittedly, there is still a great deal of speculation about some of the key events, especially the move to bipedalism. The problem is not a lack of hypotheses but an abundance. But the more detail one has about the actual evolution, the more precise and controlled can be the causal hypotheses. Compared to what we knew fifty years ago, there has been a complete sea change – a veritable quantum leap in understanding. This is a provocative but forward-looking area of science.

Talk

Big brains are all very well. It is what you do with them that counts. Obviously, you do not have to be able to talk in order to function well as a social animal – chimpanzees do very nicely – but equally obviously, language ability opens up huge new dimensions and possibilities. And even more obviously, it is a horrendous problem to try to work out the evolution of language, something that by its very nature leaves no hard trace. This does not mean that the task is absolutely impossible, and some progress has been made. To use a metaphor, there are two ways of going at the problem, through the software and through the hardware.

The first way is to take on language itself, and to see if it is something that could and did come about by natural selection. The starting point here is Noam Chomsky's (1957, 1966) theory of innate ("deep") structures in languages that mold and form every living language, linking such apparently disparate ones as English and Japanese. Chomsky is no Darwinian, thinking rather that the deep structure is something that came about suddenly. But linguists coming after him, notably Steven Pinker (1994), have argued with great force that the deep structure is just the kind of thing that could have been produced by selection. Not only is the product itself something that has an adaptive value and yet is jerry-built, as we expect of adaptations, but the ways in which children learn languages instinctively is also an indication of its Darwinian base. Not the least is the fact that such imprinting, if we may so call it, comes to an end at a relatively early age, and then, if language acquisition is to continue – especially learning a foreign language – a completely different part of the brain is brought into play.

What about the hardware? There are at least three issues here. First, there is the actual ability to speak – the vocal apparatus. The human mouth and throat are different from those of other animals and seem to exist in their present form precisely because of the need to speak. In particular, the larynx, at the top of the trachea, is lower than that of other animals like chimps. This placement, which seems to be at least nearly a half-million years old, makes possible the power of human speech. However, it is bought at the risk of getting food in the windpipe, because it is more readily misdirected from the esophagus. Too often, as steak house owners know, chunks of meat block off the breathing apparatus and the victim chokes. As the brain scientist Terrence Deacon (1997)

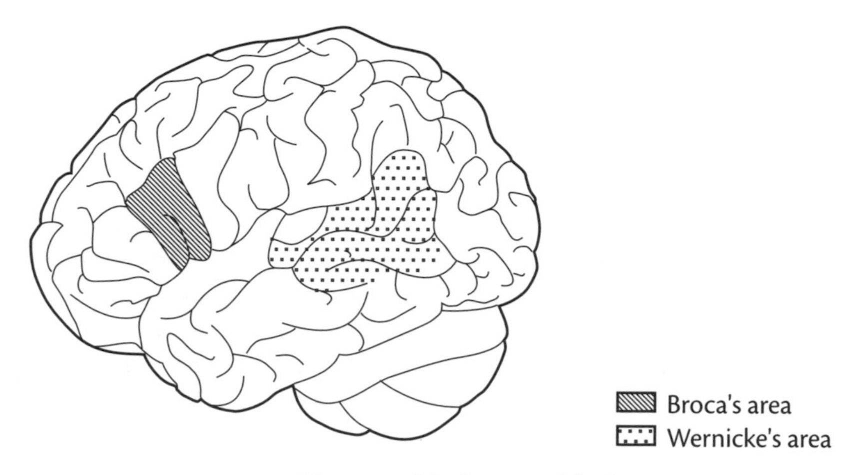

FIGURE 7.3. The parts of the brain used for language.

observes, it is hard to think that this less-than-adequate anatomy could be other than something cobbled together, Darwinian fashion, to facilitate speech. There is still some debate about whether speech as we know it began at that point. Some argue that even the Neanderthals could not speak properly (Lieberman 1984). The one firm date is that fifty thousand years ago – as inferred from the dramatic rise of culture (art, burying the dead) – we did have language. Although nothing is proven absolutely, it was probably a variant of the click language that is still spoken in parts of Africa and (uniquely) in aboriginal manhood-initiation ceremonies in Australia (Pennisi 2004).

Next, there is the evidence from the brain and the science surrounding it (Holden 2004). There are certain parts of the brain, notably Broca's and Wernicke's areas, that are crucially involved in language ability. Damage one or other of these (as in an accident or a stroke), and powers of speech decline or are ruined. The fact that these brain parts were growing even two million years ago suggests that some form of primitive proto-language probably goes back a long way (Figure 7.3). Recently, certain key genes involved in language ability have been identified and shown to be subject to natural selection. Also, there is much ongoing work on mimicry, something that is obviously a key to speech – the child learns to mimic the adult, but only in certain functional ways. Increasingly linked with this is a third approach, which looks not just at language in the spoken form, but also at gestures and physical movements.

It has been pointed out that many animals, including those close to us, communicate as much or more by sign gestures as by sounds. This leads to the suspicion that in the case of humans also, signs and gestures were (and still are) important. Children, for instance (as opposed to a bird imitator like a parrot), seem to pick up their early language as much by looking as by listening. It is possible that early language involved showing and moving, followed by lip smacking and like sounds, which then gradually turned into more sophisticated methods of making sounds. Recently very encouraging discoveries have shown that there are parts of monkey brains that seem to be involved in imitation. Even in animals like macaques that are unable to speak, the parts of the brain used for imitation correspond to Broca's area in humans.

Thought

What about consciousness itself? Certainly in the full-blooded sense, as we humans understand it, consciousness did not arrive until language evolved (Mithen 1996; Pinker 1997, 2002). But what is this thing that marks us out from others, and why do we have it? Although notoriously the French philosopher René Descartes did not think that animals are conscious – they are just machines – most people today, evolutionists or not, probably would agree that animals other than humans can be conscious. It seems hard to deny something like this to the apes, if we mean some sense of self-awareness, for instance. Apes can recognize their own reflections, and reason from there – on the basis of the reflection rubbing their faces clean of marks and the like. I suspect that most people would want to go further down the chain of being than this. Dogs and other social mammals seem to have the marks of consciousness – they can certainly show emotions like pleasure and pain, as well as pride and shame (de Waal 1982, 1996). (Observation suggests that fouling the carpet leads to shame in border collies and pride in cairn terriers.)

If you are not talking in theological terms, and identifying the mind or consciousness with the soul and insisting that only humans have souls (and hence consciousness), the right view seems to be that it is something that grows and comes into being with brain power. Ants are not conscious; dogs are, at some lower level; and adult humans truly are. More than this, as an evolutionist – as a Darwinian, particularly – one expects some kind of gradual growth or development of consciousness, very much as one has in one's own personal history. A newborn is barely conscious, if that; a philosopher is fully conscious – too much so, sometimes. It is a moot question as to how much responsibility Darwinian theory has for the explanation of consciousness. Of course, one wants to describe it and, if possible, to throw light on its nature. Certainly one wants to give a reason in terms of natural selection for its existence. But consciousness itself is just a fact, and no more. The Darwinian might well explain how important water is to life and how animals and plants use it. This can involve much discussion about the nature of water and why animals and plants do not use some other fluid like (say) mercury. But water itself and its nature – the bonds between hydrogen and oxygen – are things that the Darwinian accepts and takes for granted. Likewise with consciousness.

With these provisos, what can we say? No evolutionist questions that the explosion of brain size was essentially adaptive. For all the costs, and taking note of special circumstances like being isolated on islands – where there is both positive pressure towards conservation of resources and negative pressure from a lack of related rivals – proto-humans with bigger brains had an adaptive edge over those with smaller brains. Precisely how the brain works has always been a matter of some debate; but now, in the computer age, there are many fruitful hypotheses showing how the brain can operate as a calculator to process and use information. One particularly popular thesis invokes the idea of the brain being built on the modular pattern: there is not one central unit doing everything at once, in an all-purpose fashion, but rather units that perform different tasks, and that are connected up in various ways (Fodor 1983). The mind is like a Swiss army knife, with many tools for different purposes (although see Sterelny 2003 for a contrary view) (Figure 7.4).

What about consciousness? Even if one agrees that consciousness is in some sense connected to or emergent from the brain – and how could one deny this? – consciousness must have some biological standing in its own right. One would expect that perhaps consciousness started to emerge in a primitive way as animals developed bigger and better brains. Then it was picked up by selection in its own right and developed and refined, perhaps pulling brains along in its wake to provide the material underpinning. A lively hypothesis by the English archeologist Steven Mithen (1996) ties in the modular theory and the growth of the brain with tool use. He points out that the growth of brain size came not steadily but in two particular spurts, about two million years ago and then about a half-million years ago.

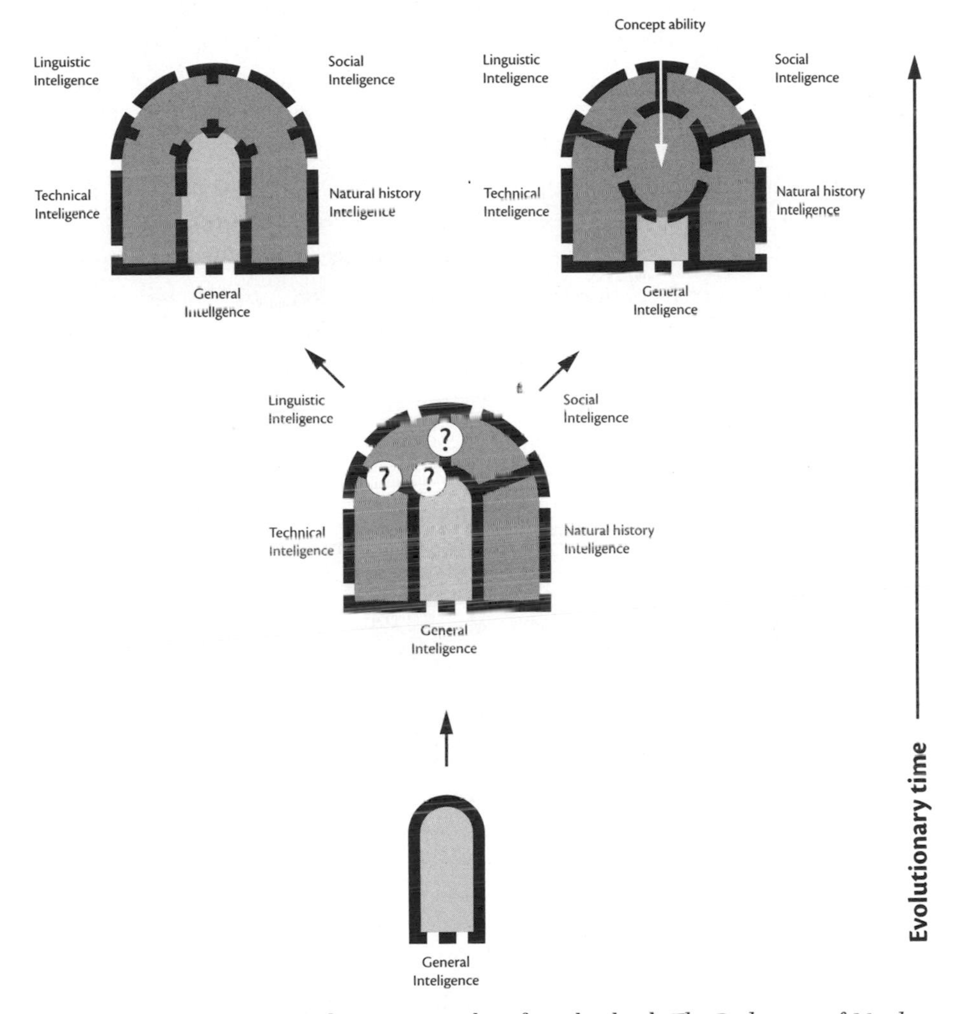

FIGURE 7.4. Illustration by Steven Mithen from his book *The Prehistory of Mind* (London and New York: Thames and Hudson, 1996). Reprinted by permission.

He suggests that before these events we had a general intelligence (possessed also by the higher apes) and then modules (also possessed by the apes) for special skills, particularly for social abilities and for navigating and understanding the environment. However, none of these were particularly well connected. With the first spurt (to *H. habilis*) came a new module for technical ability, and this is marked by the first primitive tools (chipped stones making hand axes and the like). But what is shown by the record is that there was, basically, no innovation in these tools and very limited aspects to them – there was, for instance, no use of bone and antler, despite these substances having virtues that stone does not. The second spurt (taking us to *H. sapiens*) started to bring far more integration of the various modules, so that it was then possible for sophisticated language and tool use and culture generally to take off, although even this did not really happen until (as we have seen) within the last fifty thousand years, or even later. Neanderthals were essentially left out. (They do not have elaborate burial systems or art forms or the like.)

What function does consciousness serve? Why should not an unconscious machine do everything that we can do – in which case, is consciousness just froth sitting on top of the brain's electronics? Is it a powerless epiphenomenon, to use the language of the philosophers? Almost certainly not. If something like Mithen's hypothesis is well taken, then it probably came in stages, from the awareness of animals like chimps, through a kind of nonreflective thinking (as we show when typing or riding a bicycle) of the early tool makers, to the full-blooded, reflective kind that we have. Stephen Pinker suggests that one major function is that of serving as a filter and a guide and coordinator to all of the information thrown up by the brain: consciousness sees that the brain does not get clogged up with an overload of material that it does not need and cannot use. "Information must be *routed*. Information that is always irrelevant to a kind of computation should be permanently sealed off from it. Information that is sometimes relevant and sometimes irrelevant should be accessible to a computation when it is relevant, insofar as that can be predicted in advance" (Pinker 1997, 138). Relatedly, consciousness is an important factor (part cause, part effect) in some of the very important human abilities to be discussed in later chapters of this book. It is the basis of our distinctive ways of interacting socially (where morality comes into play) and of making choices in the face of alternative possibilities (free will or choice). Consciousness gives us a power and flexibility not possessed by animals that don't have it. Mithen suggests (following the psychologist Nicholas Humphrey) that, as humans became increasingly social, consciousness was important for puzzling out the feelings and thinking of fellow humans – putting oneself in the mental shoes of others.

Even if you take the raw phenomenon of consciousness as a given, there are still legitimate questions about how mind and brain work together. The trendy position today sees mind and body as manifestations of the same substance. People of this persuasion are "monists," subscribing to the "identity theory." Here they follow Spinoza, who argued that consciousness is a manifestation in some way of material substance. This is not to say that it is just material substance – thinking is not red or hard or round – but that it is part and parcel of material substance, and nothing more is added. (Lest I sound too condescending, I should say that I am probably a monist, although I cannot say that I feel truly comfortable with any answer to the body-mind problem.) Of course, accepting an identity position is only the start of the task. One still has to say how the two-sided mind/brain functions. Perhaps here is the point where legitimately we can draw back. Consciousness is real, whether or not it is a separate thing, and it is something that seems open to the forces of evolution. More we cannot and need not say. We have to take it as a given, as of course we (as Darwinians) take the physical world as a given. Wonderful, mysterious, familiar, all of those things and more: the unexplained starting point of our inquiry.

Human Nature

What about human beings today? What would we expect to be the legacy of our evolution? The two extremes can surely be excluded at once. Simply to say that we are totally a blank slate and that our history has no effect on our present nature has to be false. We eat, sleep, defecate, copulate, and much more – all of these are animal actions, including birth and death. Simply to say that we are pawns of our genes is equally false. Think of eating and the rituals surrounding it – of how food is at the center of much social activity, or how religion impinges upon our eating, giving prohibitions and duties and obligations. The same is equally true of other human biological activities. Copulation is the most biologically natural thing we do, and yet at the same time is just about the most culturally laden thing that we do. Just think for a moment of the do's and don'ts that surround it. When did you last have sex, at midday, in the village square, with your neighbor's spouse – not using a condom?

What can we say beyond truisms? Edward O. Wilson (1978) argues that the appropriate metaphor is of a twig that is bent a bit. Culture is very powerful, but biology – our evolutionary past – shapes the way that culture develops, putting limits or constraints on its progress. Take sexuality. Some ardent feminists and others would argue that sexual desires and activities are completely a matter of culture – the way that we are brought up and the way in which we are expected and allowed to act. Some societies allow only heterosexual behavior; other societies allow and even endorse homosexual behavior and emotions. It is all a matter of the time and place (Ruse 1988b). Neither is more normal or natural than the other.

The Darwinian disagrees – and agrees! Like an Aristotelian – and the supporters of Thomistic natural law, which follows Aristotle – heterosexual activity (and presumably desire) has to be considered in some sense basic. The struggle for reproduction is the engine of evolutionary change, and human adaptations have to reflect this. This said, nothing in Darwinism suggests that homosexual behavior and inclinations are in some sense necessarily "unnatural" – in fact, the variability produced by Darwinism leads one rather to expect the contrary. Whatever the biological factors behind homosexuality – whether it is a by-product of selection for reproductive fitness (like sickle-cell anemia), or something produced in its own right (like the workers in the hymenoptera) – it could be no less a product of natural selection than heterosexuality. And if being produced by natural selection is what is meant by "natural," then so be it.

What about something like polygamy and the whole question of male–female relationships and emotions? Is it in some sense natural that men tend to be the aggressors and women hold back – "Remember, dear, even the nicest boy only wants one thing!" – or is this all a question of societal convention and expectation? The Darwinian puts humans in the overall evolutionary context – the context of mammals (primates particularly). It is a well-known fact, supported by theory and confirmed by observation, that where sexual selection is working and where there is pressure for males to acquire harems – at least to acquire multiple mates (polygyny) – you find that males are larger than females (Wilson 1975). The elephant seal is the prime – one might say, gross – example, with the males being several times the size of females. There are brutal battles between males for females, and successful males have harems of several or more females. Red deer and gorillas are also exemplars of this rule (Clutton-Brock, Guinness, and Albon 1982). Where there is less pressure toward polygyny (chimpanzees, for example), the size-related sexual dimorphism is less pronounced. In fish, where females compete for males, the trend is completely reversed, showing that this is something rooted in Darwinism and not just an ad hoc empirical generalization (Alcock 2001).

Humans are mildly dimorphic with respect to size. Some females are bigger than some males, but on average males are taller, bigger, and stronger than females (Rogers and Mukherjee 1992). This leads to an expectation of polygynous inclinations and behavior, and this is the case in most human societies. The opposite, polyandry (multiple males, one female), is very rare and tends to prove the rule. Where it occurs – Tibetan and Inuit societies are the best known – it is connected with grave difficulty in raising a family by one male unaided. Often the husbands are brothers (making the genetic relationships closer and hence less fraught if someone sleeps with one's wife), and when the pressures are released – as with Tibetans who followed the Dalai Lama and moved to India – polyandry disappears. Monogamous societies like America might be exemplifying the triumph of culture over the genes. Educate and empower women, and relationships will change. Of course, only the touchingly naïve think that America is truly monogamous – serial polygamy is a better description, and it is likely that the genes still have some major effect on these things.

Males compete for females, and again there is a general biological rule (Darwinian based) that males will be the aggressors. They provide only the sperm and then can move on – females are left raising babies, and for obvious reasons are going to be a lot more choosy, with respect to both the quality of the mate and the prospects of getting some help in child rearing (Hrdy 1999). Darwinian evolutionists think that this applies directly to humans. This explains the asymmetrical age differences often found in human relationships. Biology dictates that if you are still capable of reproduction, you will seek out partners who are likewise capable, and your biology will factor in the mate who is going to do most of the work in child rearing and the status of the one who will not. Even if a woman is not going to get much physical help, being mated to a high-status male has its virtues. You may not have approved of President Clinton's shenanigans with Monica Lewinski, but the fact that she was more than twenty years younger was no great surprise. Had she been twenty years older, you may have approved, but I expect you would have been considerably surprised.

Note that this does not mean that all males are chauvinist pigs. Given that human infants require masses of parental care, there has obviously been selective pressure – as there has been pressure in the case of birds, vertebrates that do not raise their young in hidden holes but up in trees – for males to get involved in child rearing. What a Darwinian approach to human sexuality does not imply is that all males are would-be rapists. If we have learned anything from the trends of modern evolutionary thinking, it is that the struggle is as often metaphorical as literal. In the case of humans, it is clear that the major trend has been in the direction of cooperation. We have lost our personal weapons of destruction – large teeth, for instance – and we have become comparatively mild with respect to each other. Females no longer come into heat, reducing the sexual tensions. (Imagine teaching a philosophy class in which two or three members are sending out biochemical messages that they are ready for reproduction!) Even if there might be some biological factors involved in rape – and this in itself is a highly contentious claim (not just by critics of Darwinism, but by Darwinians themselves) – there is no reason to think that it is simply a strategy unequivocally promoted by natural selection. If humans have to get on with each other, and if it is in their selective interests to do so, then forcibly copulating with the mates and relatives of others is not likely to raise one's status in the group.

Race

Now we turn to what is perhaps the most contentious of all issues to do with human nature and its biology. I live in Tallahassee, the capital of Florida. We are in the far north of the state, and often it is truly said that our culture is more south Georgia than north Florida. This certainly applies to our racial composition. Unlike Miami, the big city in the south, we have virtually no Hispanics and very few Asians. We do, however, have a population that is 40 percent African American, and anyone who thinks that this is of no significance has simply never paid us a visit. Most visibly, it shows in the geography of the city, with the north side being white and rich (some parts, very rich) and the south side being black and nonrich (some parts, very nonrich). The division runs through the city in other ways, not the least in higher education. I teach at Florida State University, which is more than 80 percent white, with many of the minority students enrolled in sports and dance and the like. We have also a "historically black university," Florida Agricultural and Mechanical University, which is almost exclusively black, white students being enrolled in just one or two professional schools.

Race matters, but is it just some kind of social construct? Is it all really a matter of culture, with biology hardly relevant? One can hardly deny that African Americans do look different from people of European stock, and one can hardly deny that this leads to differences in status and behavior and much more. In five years, I have seen fewer than five interracial (black–white) couples. But from where come the differences? Surely it cannot truly be biology? After all, most African Americans have as much European blood in their veins as African blood – as many European genes as African genes. These are emotionally loaded issues, and no answer is going to satisfy more than a minority. Some would not even have us ask the questions, because they draw attention to what they regard as fictitious entities. This is silly. Races do exist, and to say it again, race matters. The questions, rather, must be about the nature of race and then about what, if anything, we can or should do to improve things. I would improve the housing on the south side of Tallahassee. I suspect that most of the faculty and students at FAMU would tell me to leave their campus alone.

What we can say is the following. Biologically speaking, the human species is remarkable for its lack of internal genetic variation. Figures vary somewhat, but overall you are looking at about a tenth of a percent. This is what you would expect from a species that was a small population leaving Africa about 150,000 years ago. The variation that exists is spread through the species. A recent major study of human variation, looking at 1,056 individuals from 52 populations, reported: "Of 4199 alleles present more than once in the sample, 46.7% appeared in all major regions represented: Africa, Europe, the Middle East, Central/South Asia, East Asia, Oceania, and America. Only 7.4% of these 4199 alleles were exclusive to one region; region-specific alleles were usually rare, with a median relative frequency of 1.0% in their region of occurrence." Put things another way: "Within-population differences among individuals account for 93 to 95% of genetic variation; differences among major groups constitute on 3 to 5%" (Rosenberg et al. 2002, 2381).

Richard Lewontin (1982) is one who has seized on such facts as these and made them the premise for an argument concluding that there are no significant differences between races and that now the whole notion should be dropped. "Human racial classification is of no social value and is positively destructive of social and human relations. Since such racial classification is now seen to be of virtually no genetic or taxonomic significance either, no justification can be offered for its continuance." However, as

R. A. Fisher's last student, Anthony Edwards (2003) has recently pointed out, this is fallacious reasoning. If you look just at one locus, and the alleles at that locus, then there might be little genetic distance between two groups. But once you start to factor in the information gathered from one locus after another, if the variations are correlated, then the picture changes and the groups start to emerge strongly.

In the study just described, correlation does play a major factor. The differences between groups are sufficiently strong that if you run a cluster analysis across the large sample, you find that people sort into groups that correspond to ethnic sortings. Geographic Europeans come out as one genetic cluster, and Africans come out as another genetic cluster. Even more than this, as you start to factor in more and more genetic information, the divisions get ever-finer, and they continue to map ever-finer ethnic and geographical groups. And these are fairly robust findings, coming out much the same when you work from different amounts and samples of genetic information. For instance, the analysis picks out as anomalous a group in northern Pakistan. These are the somewhat isolated Kalash, who are believed not to be of the same ethnic background as the rest of their countrymen, but to have a European or Middle Eastern origin. In short: "Genetic clusters often corresponded closely to predefined regional or population groups or to collections of geographically and linguistically similar populations" (p. 2384).

What, if anything, does this information mean? At the most obvious level it means that, just as one would expect, after the human species split up, the different populations went their own evolutionary ways. Not that far, but then they have not had much time to travel – and perhaps today, with racial mixing thanks to modern methods of travel and so forth, the separate routes are coming to their ends. Are the differences simply random, or do they show the effects of selection? Probably some drift was involved – the founder principle may have been active – but selection also had its part to play. The significance of white skin compared to black is much debated – probably something to do with optimizing vitamin production given different amounts of sunshine – but selection was obviously at work in this direction, making those in northern climes lighter than those close to the equator (Jablonski and Chaplin 2003). This, incidentally, is a general pattern found in other animals. Other selection-produced features were probably body weight and size, and such things as the relatively flat facial features of Asians – again a pattern found in other

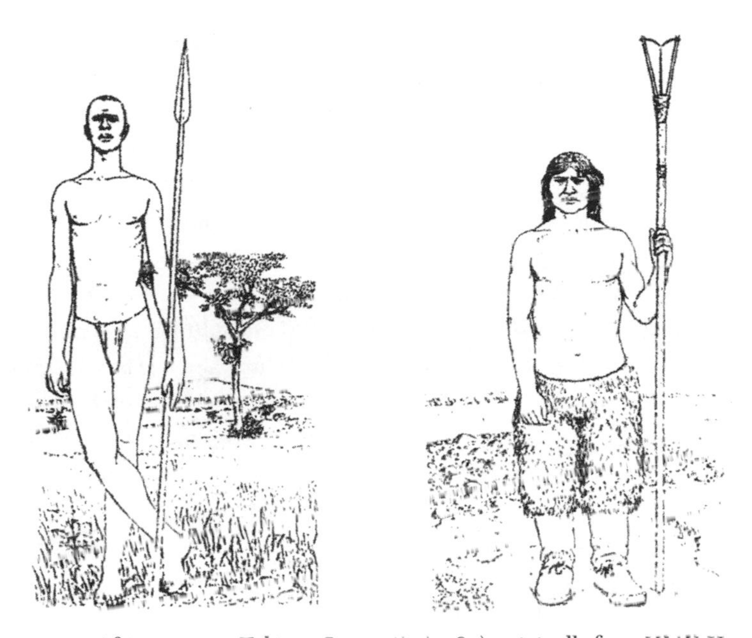

FIGURE 7.5. African versus Eskimo. Lewontin (1982), originally from W. W. Howells, "The distribution of man," *Scientific American* 203 (3). 112–27. Reprinted by permission.

animals, and probably connected to efficient ways of preventing bodily heat loss (Lewontin 1982) (Figure 7.5).

Some have suggested that, through a kind of multiplier effect, different genetic backgrounds can make for significant social and behavioral patterns. Wilson (1978) suggests that Asians are less able to metabolize alcohol than are Europeans and that this leads to different cultural habits where food and drink are being consumed. What is generally agreed is that no genetic differences swamp the effects of culture, and that (to take the most contentious issue of all) one needs to regard very carefully claims that racial differences correspond to differences in intellectual ability. In the nineteenth century, people like Herbert Spencer were convinced that people in the north had had to work harder than people in the south, and hence selection or Lamarckian inheritance of acquired characteristics had led to bigger brains and intelligence in the north (white) over the south (black). But then, in the nineteenth century people like Herbert Spencer were convinced that masturbation led to insanity. Tracking a giraffe seems no less tricky than tracking a reindeer. What is clear is that (even if there are differences in brain size) there is no ready connection with intelligence. Women are smaller than men, but no one who teaches in a North American university today would claim that women are generally less intelligent than men. What is also clear is that social factors are very significant. Again, women show the way. In the nineteenth century, even sympathetic men like T. H. Huxley did not think that generally women were up to higher education.

Let us leave things with the reflection that while the Darwinian would not deny that there might be genetically linked cultural and behavioral differences and abilities between races, this is not a necessary inference. Nor is it necessary to conclude that one race might be more talented intellectually than another. It would be cowardly to deny that this could be the case, just as it would be cowardly to deny that even though males and females might compete evenly, there might be differences, either because of selection for different roles or as side-effects of other (physical or physiological) differences. But in the case of race, until we in Florida make a genuine effort to see that all children start off with a level playing field, I am not going to make claims in this direction. And in the case of sex, one would have to be incredibly insensitive not to see the social handicaps put in the way of women, in – especially in – academia. Whatever else, as always, culture and education and opportunity will be very significant.

Reductionism

Let us suppose we were talking about another species – say, dogs. Everyone now would be feeling pretty satisfied. We would be well on the way to sorting out dogs from wolves and foxes and so forth – digging back into the fossil record, making serious hypotheses about the reasons for the behavior, and so forth. At the same time, there would be much interest in such things as dog habits and structures and mating procedures. No one would say that we knew everything. Certainly no one would say that everything to do with dogs was a function of the genes alone. But overall, the relevance of Darwinism would be beyond doubt. Why should we not say the same for humans?

Many Darwinians would say that we can say the same for humans. We do not know everything. Probably there will always be unanswered questions – although "never" is a long time. But we have learned a great deal, and much of this knowledge has come in the last fifty years. Why then be pessimistic about the future understanding of the evolution of humankind? Undoubtedly, much of the opposition is religiously based. Many in the Judaeo-Christian tradition fear that making us humans part of the evolved world in some sense threatens our special status and relationship with God. We shall raise these sorts of issues and speak to them later. Frankly, I do not see that the fears are entirely without warrant or misplaced. Already the pundits are weighing in with conclusions about *Homo floresiensis*, which are as portentously pompous as they are conspicuously callow: "If it turns out that the diversity of human beings was always high, remained high until very recently and might not be entirely extinguished, we are entitled to question the security of some of our deepest beliefs. Will the real image of God please stand up?" (Gee 2003)

But there is more at issue than this. Secular thinkers like Richard Lewontin (1991) and Stephen Jay Gould (1981) are desperately worried about many aspects of the Darwinian approach to human nature. They fear that it is reductionistic and (relatedly) deterministic. Why should this be a worry? The issue is not so much ontological or theoretical reduction, but methodological reduction: genes good, molecules better. Many of these critics have or had Marxist sympathies, and this at once sets them against methodological reductionism. Friedrich Engels (1964) argued that one of the fundamental laws of nature is that of "quantity into quality." As you go up the scale – including the size scale – you get real changes of a kind that cannot be explained in terms of lower-level entities. However, since Engels's examples tended to be banal – water boiling, for instance – and not obviously true, we can leave that line of thought, while agreeing that there is no absolute reason why explanation might always require understanding in terms of the smaller. Group selection might in some cases be preferred over individual selection, and macroevolution might in respects be inexplicable in terms of microevolution. But, as a general policy, methodological reduction works. Mendelian genes were good, molecular genes are better. We have better understanding – better understanding in terms of the epistemic principles of good science – with molecular genes than with Mendelian genes. It is that simple.

Or perhaps not that simple. Turn from epistemological questions to moral questions. Could it not be that a biological analysis (half-humorous, completely serious) of the Clinton–Lewinski relationship is on the road to exonerating him because, after all, that is "human nature"? Conversely, is one not labeling her as pretty silly – especially in her belief that the president would leave his wife for her – and again seeing this is a function of "human nature"? More seriously, if one thinks that perhaps rape does have some biological underpinnings, then one is on the way to excusing it (pro: Thornhill and Palmer 2001; contra: Travis 2003). It is all rather like trying to understand Hitler's motivations (and why some will have no part of this). To understand is to forgive – especially in the Darwinian case, because reduction is another term for determinism (Ruse 1988b). One is in effect saying that seducers and rapists are genetically determined to do what they do and hence not morally responsible. Likewise with issues of race. Say what you like, once you let the genetic genie out of the bottle, you are on the way to blaming Tallahassee's south siders' situation on their genes, and next you will be saying that there is nothing that anyone can do about it.

These are serious arguments, and only the insensitive would think that they have no basis whatsoever. Too often, bad behavior is excused as "human nature" – boys will be boys. Blacks are down the social scale because they are IQ-challenged rather than because whites have exploited them cruelly for three hundred years. But there is another side to the story. Morality can be fought with morality, and one can argue that there are moral reasons for pushing the Darwinian (methodologically reductionistic) approach. Take the case of infanticide, a topic studied by the Canadian researchers Martin Daly and Margo Wilson (1988, 1998). They have shown that a huge amount of infanticide results from step-parents (usually step-fathers) beating up the children of their spouses. It is a very well-documented fact in the animal world that males often kill the already-existing offspring of a new mate, thereby freeing her to devote exclusive attention to their own progeny. Humans are not lemmings or lions – two species where such killings are routine – but apparently something similar occurs: "having a step-parent is the most powerful risk factor for severe child maltreatment yet discovered" (Daly and Wilson 1998, cover) (Figure 7.6).

A child is over one hundred times more likely to be killed by a step-parent than by a biological parent.

Of the 87,879 maltreatment victims identified by the American Humane Association in 1976, 15% lived with substitute parents. Less than one-third of these victims exhibited overt nonaccidental injuries, and when we consider only this subset, the percentage living with substitute parents rises to 25%. Finally, we can confine attention to the most extremely and unequivocally abused children: the 279

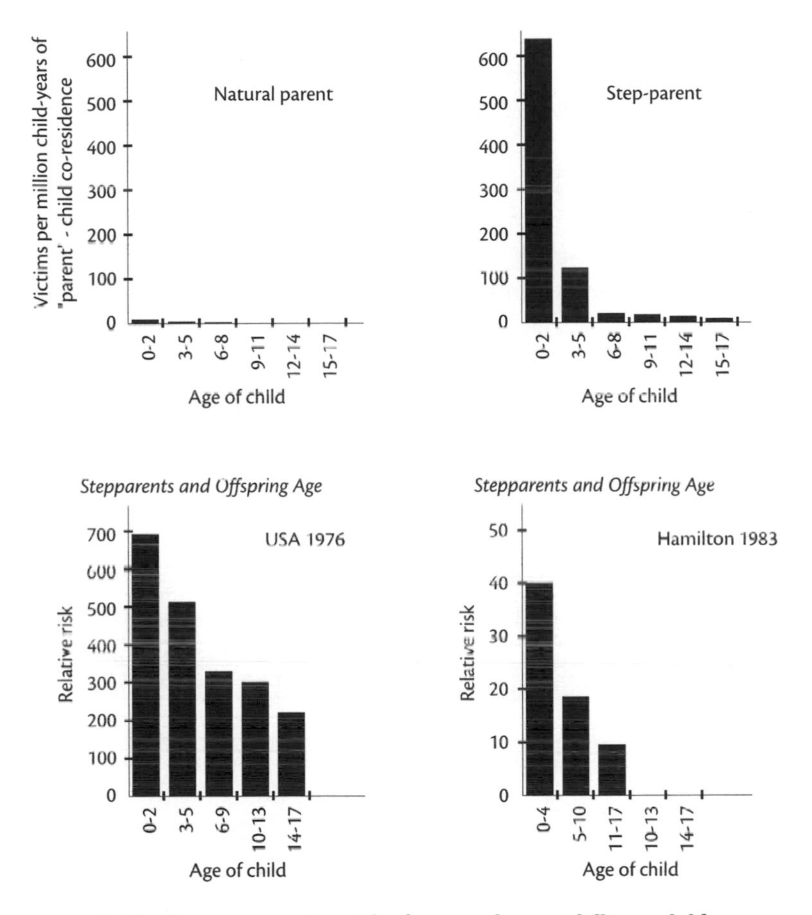

FIGURE 7.6. Top line: comparative risk of a natural parent killing a child as opposed to a step-parent. Bottom line: comparison between the United States and Canada (the Ontario city of Hamilton). From Martin Daly and Margo Wilson, *Homicide* (Hawthorne, N.Y.: Aline de Gruyter, 1988). Reprinted by permission of Transaction Publishers, Piscataway, N.J.

fatalities. Not only should reporting bias be minimal for this group, but their relative youth (median age = 3.6 years) is an additional reason to expect fewer stepparents; nevertheless, the fraction dwelling with substitute parents rises again to 43%.

Comparing this to

a random sample of American children in 1976, with an age distribution corresponding to that of the 279 fatal abuse victims, the best available national survey (Bachrach, 1983) indicates that only about 1% or fewer would be expected to have dwelt with a substitute parent. An American child living with one or more substitute parents in 1976 was therefore approximately 100 times as likely to be fatally abused as a child living with natural parents only. (Daly and Wilson 1988, 88–9)

Daly and Wilson emphasize that most step-parents do not kill their step-children. But, proportionately, a huge number do. Moreover, this is a ratio that seems to hold across cultures. The figures in Canada are comparable.

Now, what does one say here? The Australian philosopher Kim Sterelny (2003) argues that there is no reason to think that evolutionary biology has anything to do with the situation, and that step-parental abuse is very likely "generated by the failure of attachment mechanisms which normally damp down the frustrations and angers caused by dependency" (p. 104). This, however, seems to be a classic case of confusing proximate causes with ultimate causes. Even if Sterelny is right, and he may be, it does not preclude an evolutionary explanation of the reasons for the failures. It is true that if there is no evolutionary explanation, then one can just deal with such abuse in its own right. But if there is an evolutionary explanation – and for the sake of argument let us assume that Daly and Wilson are onto something – then there does seem to be more tension. Does one simply excuse the murderous step-fathers because they got into a bit of assault and battery that went too far? Because it was "human nature"? Because their genes made them do it? Most certainly not.

First of all, Daly and Wilson do not arguing that every aspect of homicide is directly and exclusively a function of the genes. Indeed, for some phenomena, cultural, nonbiological factors are surely the major causal triggers. For instance, the murder rate in the United States is over four times that of its neighbor Canada, something that is a function of many things. Undoubtedly, very significant is the ready availability of guns in one country but not the other. Second, Daly and Wilson are not arguing that every step-father kills his mate's kids. Most – the big majority – do not, and many are very loving social parents. What they are saying is that infanticide is a serious issue and that we should take note of it, try to understand it, and then do something about it. If the understanding involves biology, then so be it.

One of the most remarkable facts is that, before Daly and Wilson went looking and asking uncomfortable questions – went looking and asking questions precisely because they had an evolutionary hypothesis – a lot of places simply did not collect the pertinent data, because they did not want to upset members of reconstituted families. That has changed, and who would not say for the better? Now we know what is going on, and we have some reason why. But notice that no one is locked into some kind of rigid reductionistic determinism. What we need to know is why the majority can handle the situation, sometimes very well indeed, and why the minority fails. Then we are on the way to offering solutions to a terrible problem.

Likewise with race, it is not all cut and dried. The urological group to which I go is forever using the local paper to advertise prostate screenings – these advertisements are aimed directly and explicitly at African American men, who are a higher-risk group than others. Is one to stop this practice, in the name of ideological purity? I think not. Taking race seriously can lead to the good as well as well as to the bad. The critics are right. When dealing with human nature, there is a very fine line between the epistemological (knowledge) and ethics (morality). Darwinian approaches to humankind certainly cross that line. But this is not to say that they lead straight to the morally indefensible. People of good will can see that good as well as ill can come from such efforts. Modern science, Darwinian science, is reductionistic. This does not mean that it is thereby immoral.

CHAPTER EIGHT

Fact or Fiction?

Darwin's theories were conditioned by the patriarchal culture in which they were elaborated. . . . The *Origin* provided a mechanism for converting culturally entrenched ideas of female hierarchy into permanent, biologically determined, sexual hierarchy.

F. Erskine (1995)

This is but one little shower in a long, wet, windy season of like comments. In recent years, Darwinism has become the whipping boy of every disgruntled member of society – the root cause of problems from fascism to sexism, from anti-Semitism to capitalism, from cross dressing (I kid you not) to the breakdown of the traditional family. It is argued that Darwinism is no true scientific theory but a mere reflection of the more offensive elements of the society within which it finds itself, at best a social construction – an epiphenomenon of the culture within which it resides – and at worst a secular-religious rival to Christianity – a world picture complete with moral directives and eschatology. What is fascinating is the extent to which people at opposite ends of the social/political/religious scale join forces in their dislike. Phillip Johnson (1995, 1997), a well-known Christian Creationist, is an archetypical case of one who worries about the ways in which Darwinism corrodes the traditional values. Richard Lewontin (1991), Marxist and atheist, is an archetypical case of one who worries about the ways in which Darwinism supports the traditional values.

Progress

Let us bite the bullet. There is considerable truth in the charges. Darwinism in particular, evolution in general, has always been used as much more than a scientific theory – that has been the attraction for many people (Ruse 1996b, 2005b). Go back to the beginnings of evolutionary theorizing in the eighteenth century, and go in more detail over some of the points covered in Chapter 1. This was the age of progress, the belief that the fate of humankind lies in our own hands and not in God's, and that through effort and intelligence we can rise from the depths and scale the heights (Bury 1920; Wagar 1972). The early promoters of the idea of historical organic development were explicit in their belief that the evolution of plants and animals was but part of a larger picture, namely, that of improvement from the lesser to the greater – progress in society, progress in education, progress in science, and then progress in life forms. Evolutionary thinking was a social construction – less something reflective of the facts of nature and more something bound up with the cultural aspirations of those who promoted it.

Consider the French *philosophe* Denis Diderot, mentioned earlier, one of the first to fly the idea of organic transformation.

> Just as in the animal and vegetable kingdoms, an individual begins, so to speak, grows, subsists, decays and passes away, could it not be the same with the whole species? . . . would not the philosopher, left free to speculate, suspect that animality had from all eternity its particular elements scattered in and mingled with the mass of matter; that it has happened to these elements to reunite, because it was possible for this to be done; that the embryo formed from these elements had passed through an infinity of different organizations and developments . . . that it has perhaps still other developments to undergo, and other increases to taken on, which are unknown to us; that it has had or will have a stationary condition; . . . that it will disappear for ever from nature, or rather it will continue to exist in it, but in a form, and with faculties, quite different from those observed in it at this moment of time. (Diderot 1943, 48)

This kind of thinking did not come out of thin air. Diderot was reflecting his thought about societies and their growth from the simple to the complex, from the savage to the Frenchman. "The Tahitian is at a primary stage in the development of the world, the European is at its old age. The interval separating us is greater than that between the new-born child and the decrepit old man" (Diderot 1943, 152). What happens in society is what happens in the world of animals and plants.

Everybody else who promoted evolution felt exactly the same way, and everyone was explicit in the belief that progress – cultural and biological – was the key to everything. This was the creed of Erasmus Darwin in England, of Jean Baptiste de Lamarck in France, and of the various *Naturphilosophen* in Germany. It was also the target of those who hated evolution. Do not think that the big worry about evolution was its denial of the literal truth of Genesis. By the beginning of the nineteenth century, sophisticated thinkers were quite ready to interpret the Bible metaphorically. It was the ideology of progress that they found really threatening: in part because of its religious implications – now we are playing God and downgrading the significance of the Creator – in part because of its social implications. With reason, many saw progress as one of the radical ideas leading to the French Revolution and the Terror that it had unleashed. So people like the great French anatomist Georges Cuvier – a lifelong opponent of organic change – went after the underlying philosophy of progress as much as they did such ostensible problems as the inadequate fossil record.

This pattern continued through to the time of Darwin. Consider the anonymous author (now known to have been the Scottish publisher Robert Chambers) of the notorious evolutionary tract *Vestiges of the Natural History of Creation*, published fifteen years before the *Origin* in 1844. He makes no bones about his seeing evolution as an outgrowth of the philosophy of progress.

> The question whether the human race will ever advance far beyond its present position in intellect and morals, is one which has engaged much attention. Judging from the past, we cannot reasonably doubt that great advances are yet to be made; but if the principle of development be admitted, these are certain, whatever may be the space of time required for their realization. A progression resembling development may be traced in human nature, both in the individual and in large groups of men.... Now all of this is in conformity with what we have seen of the progress of organic creation. It seems but the minute hand of a watch, of which the hour hand is the transition from species to species. Knowing what we do of that latter transition, the possibility of a decided and general retrogression of the highest species towards a meaner type is scarce admissible, but a forward movement seems anything but unlikely. (Chambers 1846, 400–2)

Darwin and Progress

What about Charles Darwin and the theory of the *Origin*? Darwin put a crimp in this happy progressionist thinking. On the one hand, he wanted to stay away from the excesses, precisely because so many did think that evolution was nothing but an excuse for a radical philosophy. On the other hand, he could see the relativism in his mechanism of natural selection – success in the struggle and consequent adaptations do not necessarily reflect the nicest or best or highest. What works is what succeeds, and what succeeds could as well be degenerate as improved. Having said this, Darwin was as committed to the philosophy of progress as were his contemporaries, and he certainly tied it in with evolution. He had no doubt that we humans – the English especially – are at the top of the tree of life. This comes through again and again in the *Origin*, especially in the later editions that he produced between 1859, when the book was first published, and 1872, when the sixth and final revisions and additions were made.

Darwin's gambit in support of progress centered on a version of what today's evolutionists call an "arms race": as one line improves, another improves in tandem, to keep up. The prey gets faster, and so the predator must get faster. Darwin thought that overall this kind of competition would promote brains, and their possessors. The following passage was added to the third edition of the *Origin* of 1861.

> If we take as the standard of high organisation, the amount of differentiation and specialization of the several organs in each being when adult (and this will include the advancement of the brain for intellectual purposes), natural selection clearly leads towards this standard: for all physiologists admit that the specialization of organs, inasmuch as in this state they perform their functions better, is an advantage to each being; and hence the accumulation of variations tending towards specialisation is within the scope of natural selection. (Darwin 1959, 222)

After the *Origin*, especially with the lack of enthusiasm for natural selection, progress and evolution continued as tightly linked as ever before. In Chapter 1, we were introduced to the thinking of he who gave biological backing to the "White Man's Burden" – Herbert Spencer (1857), self-taught philosopher, biologist, psychologist, and sociologist. Immensely popular, he preached progress and evolution on every possible occasion. Remember that for him the essence of progress lay in a move from the simple to the complex, from what he called the homogeneous to the heterogeneous. Although the idea was absorbed with something of a French spin on things, this was the major influence on the young Henri Bergson. But even if Darwin did not dent the enthusiasm for progress, you might think that the coming of Mendelism would put pressure on the link between evolution and progress. The key fact about the raw stuff of evolution – Mendelian mutations – is that they are random. Not uncaused, but not appearing in response to the needs of their possessors. Surely this puts an end to hopes of an upward organic rise from the simple to the complex, from (as they liked to put it) the monad to the man?

Not one bit! People went right on, right through the twentieth century, promoting progress and evolution. Thus Julian Huxley, the biologist grandson of Thomas Henry Huxley:

> Evolution, from cosmic star-dust to human society, is a comprehensive and continuous process. It transforms the world-stuff, if I may use a term which includes the potentialities of mind as well as those of matter. It is creative, in the sense that during the process new and more complex levels of organization are progressively attained, and new possibilities are thus opened up to the universal world-stuff. (Huxley 1942, 131)

Most recently we have Edward O. Wilson, great admirer of Spencer and ardent evolutionary (and social) progressionist. He writes:

> [T]he overall average across the history of life has moved from the simple and few to the more complex and numerous. During the past billion years, animals as a whole evolved upward in body size, feeding and defensive techniques, brain and behavioral complexity, social organization, and precision of environmental control – in each case farther from the nonliving state than their simpler antecedents did. (Wilson 1992, 187)

Adding: "Progress, then, is a property of the evolution of life as a whole by almost any conceivable intuitive standard, including the acquisition of goals and intentions in the behavior of animals."

Even botanists add to this refrain. A leading plant evolutionist writes as follows: " 'Biological success' is not easy to measure in an unambiguous way, but by almost any yardstick the angiosperms appear to be the most successful seed plant group" (Niklas 1997, 204). Consider living angiosperms against other surviving plant groups – bryophytes, pteridophytes, and gymnosperms. There are 220,000 species of angiosperms, compared to 22,400 species of bryophytes, 9,000 of pterigophytes, and 750 of gymnosperms. The bean family alone has 14,000 species.

Yet another yardstick of biological success is ecological diversity. Here too the flowering plants are the most successful. Aside from occupying polar regions, deserts, and bodies of freshwater, the angiosperms have also returned to the sea (e.g. the "seagrass," *Zostera*), a habitat totally devoid of any bryophyte, pteridophyte, or gymnosperm species. Yet another gauge of evolutionary success is how rapidly one plant group assumes dominance over previously successful ones. The fossil record shows that the angiosperms rose to taxonomic dominance over their gymnosperm contemporaries within 40 million years or so. (pp. 204–5)

Read In or Read Out?

I am arguing that from the beginning, right down to the present, many people have regarded evolution as a kind of biological equivalent to social progress. In this respect, it has been and still is an epiphenomenon on culture. Can this really be so? Perhaps I am missing the obvious: evolution really is progressive! People see progress in evolution because it is there, objectively, in the real world. Care must be taken with this claim. As noted earlier, no one wants to invoke teleological, directed forces leading to humans – or to angiosperms, the favored food of humans. Those forces have no place in science. If there be progress, it must be the result of natural forces. Perhaps, however, far from being antithetical to progress, natural selection can and does produce it. Could Darwin have been right to focus on arms races? Julian Huxley was one enthusiast for this kind of process:

The evolution of the ungulates is not adapted merely to greater efficiency in securing and digesting grass and leaves. It did not take place in a biological vacuum, but in a world inhabited, *inter alia*, by carnivores. Accordingly, a large part of ungulate adaptation is relative to the fact of carnivorous enemies. This applies to their speed, and, in the case of the ruminants, to the elaborate arrangements for chewing the cud, permitting the food to be bolted in haste and chewed at leisure in safety. The relation between predator and prey in evolution is somewhat like that between methods of attack and defence in the evolution of war. (Huxley 1942, 495–6)

Taking this kind of thinking to its logical fulfillment, Richard Dawkins (1986) argues that arms races in nature have gone the way of military arms races, from clashes between armor plate and guns to the development of sophisticated electronic methods of attack and defense, from clashes between shells and boring mechanisms to the development of ever-more-sophisticated on-board computers, otherwise known as brains. In other words, the kind of relativistic progress of particular arms races adds up eventually to a kind of absolute progress, with humans coming out on top.

Seductive although all of this is, however, it really will not do (McShea 1991). On the one hand, there are serious questions about how prevalent or effective arms races really are in nature. On the basis of the fossil evidence, it has been argued that the predator/prey increase-in-speed hypothesis does not find universal confirmation (Bakker 1983). For long periods of time, there have been many cases where a maximum speed seems to have been reached and stabilized out. So although arms races may be important, they are probably not all-important. On the other hand, and even more importantly, even if arms races are all-powerful and prevalent, there is no obvious reason to think that (in the animal world) they will lead to humans. *Homo floresiensis* shows that evolution does not necessarily end with big brains – they apparently are the descendents of creatures with significantly larger brains than they had. To reiterate a point made in the last chapter, brains are very expensive items to maintain. As guerrilla warriors show us only too painfully today, often one can do more in war with light and relatively inexpensive weapons than with high-priced, complex equipment.

Stephen Jay Gould put things neatly. "Since dinosaurs were not moving toward markedly larger brains, and since such a prospect may lie outside the capabilities of reptilian design . . . , we must assume that consciousness would not have evolved on our planet if a cosmic catastrophe had not claimed the dinosaurs as victims. In an entirely literal sense, we owe our existence, as large and reasoning mammals, to our lucky stars" (Gould 1989, 318).

Niche Climbing?

Recently a somewhat different attempt has been made to get progress out of the Darwinian evolutionary process. This is due to the Cambridge paleontologist Simon Conway Morris (2003). He invites us to think anew about the nature of evolution and the way that it works. Although he is a deeply committed Darwinian, he argues that constraints are going to have a major effect on the course of evolution. Only some areas of potential morphological space are going to be able to support functioning life. To use an example of our own, encountered earlier, while you obviously do have a space able to support cats of the size that actually exist, you do not have a space able to support cats of an elephantine size. The flip side to this – and here the Darwinism comes to the fore – is that natural selection is constantly pushing organisms to find new areas to colonize, and so if something is an area of potential morphological space able to support life, it will be invaded. Indeed, we might expect it to be invaded many times. Conway Morris points out that this is not just an a priori inference, but something confirmed again and again in real life. There are repeated examples of convergence, of organisms of very different backgrounds and ancestries nevertheless finding their way into almost identical ecological niches (or rather, the same niche), with almost identical adaptations to help them survive and succeed in such niches. The most famous example is that of the saber-toothed tiger type animals. There were North American placental mammals (real cats) matched by South American marsupials (thylacosmilids) – both were predators, with similar behaviors and abilities, whose formidable weapons were massive teeth that could be used for shearing and stabbing. Natural selection found the open niche twice and filled it both times. Indeed, there is evidence that the dinosaurs may even have found this niche before the mammals did.

It is fairly easy now to see how a notion of progress can be extracted from or built upon this kind of thinking – how we can get the nigh-inevitable emergence of humans or at least of human-like beings, with big brains that support superior intelligence ("humanoids," as such beings have been called). We know that a niche exists for beings with big brains and superior intelligence – we humans have found it and invaded! More than this, we did so because of the forces of natural selection, and for no other reason. Hence, it is reasonable to think that it might happen all over again.

> If brains can get big independently and provide a neural machine capable of handling a highly complex environment, then perhaps there are other parallels, other convergences that drive some groups towards complexity. Could the story of sensory perception be one clue that, given time, evolution will inevitably lead not only to the emergence of such properties as intelligence, but also to other complexities, such as, say, agriculture and culture, that we tend to regard as the prerogative of the human? We may be unique, but paradoxically those properties that define our uniqueness can still be inherent in the evolutionary process. In other words, if we humans had not evolved then something more-or-less identical would have emerged sooner or later. (Conway Morris 2003, 196)

This is an ingenious line of argument. Whether it gets Conway Morris quite what he wants is another matter. Perhaps the argument does show that something human-like was bound to (or very likely to) evolve sooner or later. It is true that at some level it is assumed that niches are things out there, waiting to be occupied, rather than things that are created by organisms and that, given another evolutionary story, would not exist at all. Obviously this assumption cannot be entirely true (Sterelny and Griffiths 1999). If there were no trees, then there would be no canopies providing niches for monkeys and insects and birds and so forth. And, granting the preexistence of niches, independent of the forms into which organisms have evolved, one still does not have the conclusion that humans are superior. They are different, but we knew that already. Progress demands that in some sense they be absolutely better. And this Conway Morris has not shown.

Nor, for the sorts of reasons given already in this chapter, is it clear how he or anyone else could ever show that natural selection leads to absolute progress. Conway Morris seems to think that there is a natural progression – water, earth, air, culture. But why is water, for instance, less than the others – from a biological perspective? Cetacea (whales and other marine mammals) went back to the sea, from being hippopotamus-like creatures. This required considerable simplification of the skeleton, including the shrinking of the limbs. But whales are no less efficient than hippopotami – they have all sorts of fancy adaptations to sea life, including the ability to dive deep and stay down there. Is this degeneration or progress? Put the question in human terms. If *Homo floresiensis* had survived and *Homo sapiens* had gone extinct, what price progress then?

Moral Directives

Seeing evolution (however powered) as a progressive process leading up to humankind is a human-created picture rather than a disinterested reflection of objective reality. The vision is redolent with human culture – a social construction, if you will. In Chapter 1, I spoke of the progress-based vision of organic history as the foundation of a kind of secular religion, meaning something that functions as a kind of myth, that gives meaning to existence and that explains the special status of humans. Was that really true, and is it still true today? In a way, it is, because we have a picture of origins, with a special place for humankind as the climax. Wilson, who is an intensely religious person and who believes that religion must always play a significant role in human affairs, is unambiguous. Darwinism equals materialism equals Christianity rival.

> But make no mistake about the power of scientific materialism. It presents the human mind with an alternative mythology that until now has always, point for point in zones of conflict, defeated traditional religion. Its narrative form is the epic: the evolution of the universe from the big bang of fifteen years ago through the origin of the elements and celestial bodies to the beginnings of life on earth. The evolutionary epic is mythology in the sense that the laws it adduces here and now are believed but can never be definitively proved to form a cause-and-effect continuum from physics to the social sciences, from this world to all other worlds in the visible universe, and backward through time to the beginning of the universe. Every part of existence is considered to be obedient to physical laws requiring no external control. The scientist's devotion to parsimony in explanation excludes the divine spirit and other extraneous agents. Most importantly, we have come to the crucial stage in the history of biology when religion itself is subject to the explanations of the natural sciences. As I have tried to show, sociobiology can account for the very origin of mythology by the principle of natural selection acting on the genetically evolving material structure of the human brain.
>
> If this interpretation is correct, the final decisive edge enjoyed by scientific naturalism will come from its capacity to explain traditional religion, its chief competition, as a wholly material phenomenon. Theology is not likely to survive as an independent intellectual discipline. (Wilson 1978, 192)

Religions tend to have moral directives – "Love your neighbor as yourself" and that sort of thing. Does this kind of (let us call it) Darwinizing have a set of moral directives? Historically, this was certainly true – it was what became known as "Social Darwinism" (Bannister 1979; Richards 1987). Supposedly, you take the process of evolution, you add a couple of sentiments of obligation, and there you have a full-blown ethical theory. Despite the name, Darwin himself had little to do with the main thrust of this movement. The leadership role was taken by Herbert Spencer, who seems to be a paradigmatic exemplar of this move into evolutionary ethicizing. You start with the struggle for existence in the living world, you transfer it directly to the human world of culture and society, and you come out with a doctrine that promotes unimpeded struggle and success to the winners: laissez-faire economics. Thus Spencer wrote:

> We must call those spurious philanthropists, who, to prevent present misery, would entail greater misery upon future generations. All defenders of a Poor

Law must, however, be classed among such. That rigorous necessity which, when allowed to act on them, becomes so sharp a spur to the lazy and so strong a bridle to the random, these pauper's friends would repeal, because of the wailing it here and there produces. Blind to the fact that under the natural order of things, society is constantly excreting its unhealthy, imbecile, slow, vacillating, faithless members, these unthinking, though well-meaning, men advocate an interference which not only stops the purifying process but even increases the vitiation – absolutely encourages the multiplication of the reckless and incompetent by offering them an unfailing provision, and *discourages* the multiplication of the competent and provident by heightening the prospective difficulty of maintaining a family. (Spencer 1851, 323–4)

You cannot be more blunt than this, although as a matter of historical accuracy, in Spencer's own case the connection between his social thought and his evolutionism – both deeply rooted in progressivism – was somewhat more complex than that of straight transference. Although there are passages of the same sort after he became an evolutionist, the passage just quoted came only at the point when he was becoming an evolutionist. Moreover, although Spencer certainly believed in the struggle for existence, he saw this more as a stimulus to further work (that would push one up the chain of life) than as something that led to outright success and failure and consequent selection. This kind of thinking is to be found also in some of Spencer's later followers – the Scottish-American industrialist Andrew Carnegie, for instance. He is known today for his philanthropy in founding public libraries. The motivation was to open places where poor but bright students could go and better themselves. With Spencer, the emphasis was generally more on promoting the success of the successful than on ensuring the failure of the failures. Effort was going to improve things in a Lamarckian fashion, rather than selection in a Darwinian fashion (Russett 1976).

Conflicting Prescriptions

There were certainly those who did take Darwinism – the struggle particularly – and read it into human affairs as a justification for war and conquest. German militaristic thinkers endorsed such ideas at the beginning of the twentieth century. "Struggle is . . . a universal law of Nature, and the instinct of self-preservation which leads to struggle is acknowledged to be a natural condition of existence. 'Man is a fighter'" (von Bernhardi

1994, 13). And: "might gives the right to occupy or to conquer. Might is at once the supreme right, and the dispute as to what is right is decided by the arbitration of war. War gives a biologically just decision, since its decisions rest on the very nature of things" (p. 15). Hence: "It may be that a growing people cannot win colonies from uncivilized races, and yet the State wishes to retain the surplus population which the mother-country can no longer feed. Then the only course left is to acquire the necessary territory by war" (ibid.). This sort of stuff was picked up by an unemployed painter in the Austro-Hungarian Empire, and in the 1920s was spewed forth in the bible of the Nazi movement, *Mein Kampf*.

But just as religious people differ in their interpretations of what Jesus demands of us, so those applying evolution to social theory differed in their conclusions and moral prescriptions. For instance, countering the militaristic thinkers, the codiscoverer of natural selection, Alfred Russel Wallace (1900), was a lifelong socialist and thought (by relying on a group-selective interpretation of the evolutionary process) that nature makes us integrated and harmonious and that we therefore should strive to maximize this condition. Relatedly, the Russian Prince Petr Kropotkin (1902), an ardent anarchist, thought that evolution depends on a tendency to help each other – mutual aid – and that the right thing therefore is to abolish governments and to let our natural inclinations take over.

Generally, the enthusiasts of Social Darwinism – whether it went under that name or not – reflected the moral concerns of their day. In other words, they behaved exactly as everyone else did, including Christians and other religious people. Thus Julian Huxley, writing when the world knew too well the horrors of a huge and dreadful recession, promoted (as did Roosevelt) massive public works, like the Tennessee Valley Authority's damming of rivers and bringing of electricity to the South. He had to temper his enthusiasm, given that this was also the time of Adolph Hitler, but the message is clear.

All claims that the State has an intrinsically higher value than the individual are false. They turn out, on closer scrutiny, to be rationalizations or myths aimed at securing greater power or privilege for a limited group which controls the machinery of the State.

On the other hand the individual is meaningless in isolation, and the possibilities of development and self-realization open to him are conditioned and limited by the nature of the social organization. The individual thus has duties and responsibilities as well as rights and privileges, or if you prefer it, finds certain outlets and satisfactions (such as devotion to a cause, or participation in a joint enterprise) only in relation to the type of society in which he lives. (Huxley 1943, 138–9)

Edward O. Wilson, writing today at a time of environmental crises like the rape and destruction of the Brazilian rain forests, shows his moral concerns by arguing that humans have developed in symbiotic relationship with the rest of nature, and that we can survive only in a world of biological diversity. We have a natural inclination to what Wilson (1984) calls "biophilia." A world of plastic would kill, literally as well as metaphorically. For this reason, we must promote the preservation of such diversity. As Wilson writes in a recent book, *The Future of Life*: "a sense of genetic unity, kinship, and deep history are among the values that bond us to the living environment. They are survival mechanisms for us and our species. To conserve biological diversity is an investment in immortality" (Wilson 2002, 133).

What about the topic that opened this chapter: the nature and status of women? The conflicting messages appear here also. Some evolutionists, starting with Darwin himself, have used the theory to promote a vision of sexual nature, and some evolutionists, starting with Darwin himself, have had a vision that was less than complimentary to women. In the *Descent of Man*, Darwin wrote: "Man is more courageous, pugnacious, and energetic than woman, and has a more inventive genius" (Darwin 1871, 2, 316). Balancing this, woman has "greater tenderness and less selfishness" (p. 326). Apparently this is all part of our biology, because sexual selection has made men strong and vigorous, and women weak but selfless. Today, although he does not put things in quite these stark terms, Wilson echoes many of Darwin's sentiments, thinking that there are basic differences between men and women, and that these reflect their innate biological natures, put in place by natural selection. We may be equal but we are different, and Darwin shows the way.

However, other evolutionists, starting with Wallace, have used Darwinism to argue for a feminist vision of humankind. Wallace thought that women were going to take over the choice of sexual partners and that they would chose only the best and most sensitive types of men. Thus the one sex would be brought up to the standards of the other, higher sex.

In such a reformed society the vicious man, the man of degraded taste or of feeble intellect, will have little chance of finding a wife, and his bad qualities will die out

> with himself. The most perfect and beautiful in body and mind will, on the other hand, be most sought and therefore be most likely to marry early, the less highly endowed later, and the least gifted in any way the latest of all, and this will be the case with both sexes. (Wallace 1900, 2, 507)

A perhaps-more-realistic feminist Darwinian vision has been promoted more recently by the sociobiologist Sarah Hrdy (1981). She argues that human females have evolved to conceal their time of ovulation (unlike other animals) and that, because of this, females keep men on a string, as it were. Males do not know when or if they have impregnated females, so they must stay around with their mates, and this sucks them into child care and provision. Far from being the unwilling pawns of males, Hrdy argues that males are the pawns of manipulative females, who control relations between the sexes. Obligations stem from this asymmetry.

In the last chapter, I touched briefly on some of these issues. Without making any commitments, I admitted that it could be true that there are differences in the thinking (and emotions) of men and women. Here my intent is not to slide in my opinions under the guise of talking about others. My point is that people do draw moral and social implications from Darwinism, and that those who do often do not speak with one voice. They draw implications but of different kinds.

Objectivity

By now, the supporter of Darwinism may fear that I have conceded far too much. With all of this Darwinizing, what place is there for objective science? Why should we take Darwinism seriously at all as a reflection of objective reality, as something that shares the aims of genuine science? Simply because what I have been describing thus far in this chapter is only part of the story, the other part of which I have been describing in the earlier chapters (Ruse 1999b). Go back to the history presented in Chapter 1. Darwin's legacy split into two after the *Origin* – although this was not always clearly recognized, and still is a closed book to many today. Darwin's dream of a real science of evolution based on natural selection was one branch – this branch really starting to flourish only after about 1930, when the population geneticists brought Mendelian genetics into the picture. Darwinism as a secular religion was the other branch, and this (as we have just seen) flourished from the time of the *Origin* right down to the present day. So although there is clearly a constructivist side to the picture, it is incorrect to argue – as do critics – that this and this alone is the whole story. There is a genuine science that is not simply a reflection of the norms and values of the society within which it is embedded.

It is easy to say this, but can this really be so? Is there a nonconstructivist side to the picture? If you are not convinced by this stage of the book, probably you never will be convinced. But without being unduly pedantic, let me point out that the science presented up to this point does fit (readily and easily) the kinds of criteria that are usually associated with good, genuine, objective science. Everyone who writes on these subjects stresses that the mark of such science is that it manifests epistemic values; it conforms to rules incorporating such values (McMullin 1983). Among these values are internal coherence, external consistency, unificatory power (consilience), predictive ability and fertility, and simplicity. Work that shows these features is felt to reflect the world outside us, and work that does not show these features is thought to be a mere invention or fiction. (What about falsifiability, Karl Popper's [1959] famous criterion of demarcation between good or genuine science and bad or pseudo-science? I take it that the just-given list incorporates this criterion, particularly through consistency and predictive ability.)

Darwin's theory (and the subsequent work in his tradition) manifests these values. Some it does easily and obviously. Above all, the theory of the *Origin* was the epitome of a unifying, a consilient theory – by design. Under the umbrella of evolution through natural selection, Darwin brought in just about every area of biology – instinct, paleontology, biogeographical distribution, anatomy, embryology, classification – and showed how they are illuminated by the mechanism and in turn support the mechanism. The same is even more true today, as Darwinian scientists extend and refine their thinking. Other values have been harder to satisfy. Notoriously, the physicists of Darwin's day calculated that the Earth is far younger than seems possible as the end result of such a leisurely mechanism as natural selection (Burchfield 1975). For nearly half a century, there was great tension between biology and physics on this issue, with the biologists trying desperately to reduce the time needed for evolutionary change. This was one of the reasons why natural selection was not a great success. Finally, at the beginning of the twentieth century, it was the physicists who realized that they were wrong. They had been ignorant of the warming effects of radioactive decay, and had thus dramatically underestimated the length of Earth's history – a length that was quite sufficient for natural selection. External consistency had been achieved.

Some of the epistemic values and Darwinian biology are controversial still. Complaints are always made that Darwinism is not truly predictive. You cannot say what will be the future of the elephant's trunk or the giraffe's neck. This is true, but it is to take a too-narrow perspective on prediction. One can and does make straightforward predictions – predictions that are successful – for instance, about the ways in which bacteria and viruses will develop defenses against substances that initially are fatal to their well-being. One thinks particularly of penicillin and venereal diseases. One can also make predictions (sometimes known as retrodictions) about the ways in which natural selection acting in the past led to expected effects. Take islands. There are lots of expectations about the inhabitants of such habitats – that recent immigrants will have reduced variability, that flying organisms (especially light ones like insects) will develop adaptations (like reduced wings) to avoid being blown away, that there will be evolution to atypical size forms – the killer birds of New Zealand that occupied the niche of absent mammals, and the little people of Indonesia who hunted the little elephants of Indonesia – and much more of this ilk.

Darwinian theory is also incredibly predictively fertile. Take but one example, Edward O. Wilson (1980a, b, 1983a, b) and his work on the caste systems of leaf-cutter ants – insects that grow fungus on retrieved leaves, which they then feed to their young. Working with Adam Smith's notion of a division of labor – you can do things much more efficiently if you divide the tasks among many specialists, rather than each person trying to do things individually – Wilson was able to explain why the ants have their different castes – large soldiers to defend the nest, medium-sized leaf cutters, smaller workers to tend the fungus, and even smaller ants to tend the young. He was even able to show that the sizes and numerical proportions were exactly as one would expect if natural selection were working to optimize the efficient use of resources and labor.

Finally, a brief mention of simplicity, a value that is much prized by scientists themselves, although philosophers often have trouble explaining why it should be so prized (Hempel 1966). Natural selection itself is the paradigmatic example of a simple (not simpleminded) mechanism. As T. H. Huxley said in frustration and admiration: "How stupid not to have thought of that oneself!" More recently, my favorite example is Hamilton's explanation of the sterility of the workers in the hymenoptera, and Wilson's reaction to it – a reaction that underlines what I said earlier about the significance of the move in the early 1960s from a group-selective perspective to an individual-selective perspective. At first, the American could not believe that some unknown Englishman could have solved the mystery of hymenopteran sociality. Then Wilson came around, because such a pretty and simple idea just "could not be wrong!" Wilson describes a train journey south from Boston to Miami. It is one of the most exciting pieces of writing about science ever, so no apologies for quoting in full:

> I picked Hamilton's paper out of my briefcase somewhere north of New Haven and riffled through it impatiently. I was anxious to get to the gist of the argument and move on to something else, something more familiar and congenial. The prose was convoluted and the full-dress mathematical treatment difficult, but I understood his main point about haplodiploidy and colonial life quickly enough. My first response was negative. Impossible, I thought; this can't be right. Too simple. He must not know much about social insects. But the idea kept gnawing away at me early that afternoon, as I changed over to the Silver Meteor in New York's Pennsylvania Station. As we departed southward across the New Jersey marshes, I went through the article again, more carefully this time, looking for the fatal flaw I believed must be there. At intervals I closed my eyes and tried to think of alternative, more convincing explanations of the prevalence of hymenopteran social life and the all-female worker force. Surely I knew enough to come up with something. I had done this kind of critique before and succeeded. But nothing presented itself now. By dinnertime, as the train rumbled on into Virginia, I was growing frustrated and angry. Hamilton, whoever he was, could not have cut the Gordian knot. Anyway, there was no Gordian knot in the first place, was there? I had thought there was probably just a lot of accidental evolution and wonderful natural history. And because I modestly thought of myself as the world authority on social insects, I also thought it unlikely that anyone else could explain their origin, certainly not in one clean stroke. The next morning, as we rolled on past Waycross and Jacksonville, I thrashed about some more. By the time we reached Miami in the early afternoon, I gave up. I was a convert, and put myself in Hamilton's hands. I had undergone what historians of science call a paradigm shift. (Wilson 1994, 319–20)

Values

Have I not given the game away? Surely one of the marks of genuine epistemic science is that it is value-free – not epistemic value-free, but free of the values of society or the individual. Newton's theory does not tell us how one would like the world to be; it tells us how the world is. If Darwinism is to be considered at this level, then it too should tell us how the world is, not how we would like it to be. That was why the stuff discussed earlier in this chapter does not qualify as genuine science, precisely because it was in the prescription business. Even my example of predictive fertility fails at this point. It centered on Wilson and his ants, as he used metaphorically the Adam Smith concept of the division of labor – using it not peripherally but absolutely vitally. Yet if you were looking for a culturally value-laden notion, you simply could not find a better example than the division of labor. Smith, and those who followed him – especially including Charles Darwin, who made extensive use of the idea – thought that it was a wonderful thing, the very key to a successful Industrial Revolution. Someone like Darwin – the grandson of a man who had made his fortune in employing the notion – was not about to repudiate its worth (Ruse 1999b). Hence, inasmuch as metaphors like these get smuggled into biology – more precisely, get hauled into biology wholesale – cultural values are being brought in. Wilson may not have thought of this directly, but his work on the ants is a ringing endorsement of the division of labor – a capitalist linchpin if ever there was one.

In response, let me make two points. First, like the rest of science, evolutionary biology does use many metaphors – tree of life, struggle for existence, natural selection, arms races, and above all organisms as objects of design – and the culture of the day thereby floods and informs the science. Some would argue that the metaphors operate only initially and then get dropped as the science matures, but this is nonsense. The metaphors are in evolutionary biology and are there to stay. And a good thing, too! Without the metaphors, predictive fertility would grind to a halt. How on earth could a paleontologist explore a new fossil find without the metaphor of design? Why are the little people discovered in Indonesia little people? Why did they reduce in size, including brain size, from the forms they had been a million years ago? What selective factors were operating? None of these questions could be asked if one did not think of the little people as objects of design – of natural selection, a secular design, but metaphorically of human (or God's) design.

Second, because one uses metaphors drawn from culture, it does not follow that one endorses the values of the culture. Darwin did think that the division of labor was a good thing. Wilson may or may not think this. He may think that in the human world, the division of labor is soul-deadening – those hours spent on the assembly line screwing up widgets – and that it may not always be that efficient. As Japanese auto makers showed Detroit, it can be more efficient to have workers form teams doing all of the tasks together, to keep up the interest level. In the case of the unthinking, unfeeling ants, this is not an issue, and so the metaphor can be used – without endorsing it as a good thing for us humans. Likewise, a Quaker evolutionary biologist might find the arms race concept incredibly useful – understanding the moves and counter-moves made through evolution, like the ways in which parasitic birds (like cuckoos) and their hosts devise strategies to promote their own ends – and yet deplore war and militarism entirely.

In the light of these points, there is no good reason to think that the professional side of evolutionary biology, the professional side of modern Darwinism, is simply an excuse for promulgating the values of modern (or past) society. What then is the status of today's mature, scientific, Darwinian, evolutionary biology? One has to transcend dichotomies of objective/subjective, discovered/created, description of reality/social construction. Science, Darwinism in particular, falls on both sides of the divides. Today's evolutionary biology is clearly subjective, in that through its metaphors it reflects the culture in which it was formulated – Western (especially British and American), industrialized, mass agricultural, Christian, militaristic, sexual, and much more. We could not have had the theory had not we been living in a Judeo-Christian type of society, asking about origins and about humans and so forth. The Greeks did not ask these kinds of questions, and they were not evolutionists. The same goes for notions of struggle and selection and so forth. On Andromeda, say, there might be a society of intelligent scientists who simply don't find our kinds of questions that interesting. Perhaps some metaphors are universal – you cannot have intelligence without the ability to make artifacts, and this spells design. But, generally, the members of such an alien society would not be evolutionists; they would certainly not be Darwinians, because they would frame questions in different ways.

What then of objectivity? I am not saying that anything goes. The aliens might not be evolutionists as we understand the term, but they would not be – could not be – Creationists. It is here that the epistemic values come into play, and were the aliens to couch questions about Creationism, the epistemic values would soon show that it is no genuine science. It is inconsistent with the rest of science; it leads to no new and confirmed predictions, and so forth. The same is true down here on Earth. Darwinian biology gains objective status – it is no mere epiphenomenon of culture – because it is epistemically successful. It does what is needed to tell us in a disinterested fashion about the world of experience. It works, and that in the end is why it deserves our attention and support. Until and unless a more powerful rival appears on the scene, that is why we should be Darwinians.

CHAPTER NINE

Dishonest Science?

At the very moment that the peppered moth experiments were establishing the Oxford biologists as masters of their world, their personal and professional relationships were disintegrating in a miasma of recriminations, intrigue, jealousy, back-stabbing, and shattered dreams. They conceived the evidence that would carry the vital intellectual argument, but at its core lay flawed science, dubious methodology, and wishful thinking. Clustered around the peppered moth is a swarm of human ambitions and self-delusions shared among some of the most renowned evolutionary biologists of our era.

J. Hooper (2002)

This is but one of the milder charges made in a recent book about famous experiments of the Englishman H. D. B. Kettlewell, someone who took up the Victorian insights about melanism and the peppered moths and who tried to put the whole story on a firm basis of observation and experiment. Apparently the work was not just bad, but fundamentally shoddy to the point of dishonesty. Many, including the Berkeley-trained biologist and supporter of the Reverend Sun Myung Moon, Jonathan Wells (2000), argue that the moth work was well past the point of dishonesty, and that this is but the tip of an iceberg. From the very beginning, Darwinism has been marred by plagiarism, fraud, and charlatanism. It seems almost to attract the morally impaired. At the risk of fouling ourselves, let us scoop into this cesspool, and see what stinking objects we haul up. But as we prepare to do so, let us ask what kinds of charges might be leveled, why they are thought important, and why they might be made. Life is never quite as simple as it appears at first sight.

Plagiarism and Fraud

Like the ants, scientists are social animals. They do not work in isolation – they cannot work in isolation. There is the myth of the lonely genius in his or her garret, working by candlelight and overturning accepted conventions. But this is rarely the case. Look at Charles Darwin, who lived in the isolated village of Downe. His correspondence shows us that he had the biggest and widest range of people who were feeding him information; that he mixed with leading scientists of the day, either at home or on remarkably frequent trips to London; and that his publications were read and commented on by experts and laypeople all around the globe. Even if one did have a genius without friends, he or she could do serious work only by building on those who came before. No one picks up from scratch. Science starts with problems – the mystery of mysteries – and these are posed by others. Science ends with dissemination of results. Something discovered and not announced is not science.

For these reasons, certain transgressions are viewed very seriously by scientists (Hull 1988). One is not talking now about straightforward moral transgressions, like theft or harassment. These are wrong, but not scientific transgressions. One is talking now of things that spoil or disrupt the work of science – most particularly of plagiarism and of fraud. Plagiarism is pinching someone else's ideas or results; fraud is fabricating ideas and results. Fraud is worse than plagiarism. Scientists depend on each other, crucially. No one can spend time replicating the work of others before launching on his or her own work. Phony science means that one's own work is thereby tainted. But plagiarism is also inexcusable. The real reward in science, far more than any monetary payoff, is the respect of one's peers, earned through one's work. Darwin was clear on this. If you steal someone else's work, you are taking that which is not yours, and depriving others of their earned credit.

For all that one has some well-documented cases of plagiarism and fraud in science, generally speaking they are not as common as they might be. Why is this? Agreed that scientists are no less moral than most – they might be more moral – but the real reason digs more deeply. There is a strong aesthetic side to science – remember the enthusiasm for simplicity – and it plays a role here. Cheating is not just wrong, it breaks the rules of the game – it is akin to a perversion (like eating rotting dead animals) that no one would want to engage in even if they could. Remember Gyges in Plato's *Republic.* He had a ring that rendered him invisible and hence able to do anything without penalty. He slept with the queen and killed the king and grabbed his job. He did not do the perverse. Edward O. Wilson is a moral man, but the real reason why he did not – could not – pretend that he had discovered Hamilton's ratios himself was that it would have violated his own deepest sense of himself as a scientist. Such pretense would be akin to eating rotting dead animals.

A consequence of all of this is that scientists tend to view plagiarism and fraud as loathsome things that put the perpetrator outside the bounds of decency. One can understand someone fiddling the funds to take a lover off to an exotic island – even if one is mad that the funds are no longer available for others – but one simply cannot understand someone altering the figures on a graph, just to get a paper in *Science* or *Nature*. For this reason, if for no other, scientists tend to be very cautious about making claims of plagiarism or fraud – it simply seems impossible that a person would do it, just as it seems impossible that a person would willingly eat rotting dead animals. Especially if there are other reasons to venerate the supposed sinner.

A nice example of this concerns the father of genetics, Gregor Mendel, and his fishy figures. Mendel did his key experiments on pea plants, grown in his monastery garden, and he dutifully recorded the results that led him to suppose his ratios. Unfortunately, fifty years after Mendel's death, Ronald Fisher (1936) ran the figures through his much more sophisticated mathematical apparatus, and it turned out that Mendel's empirical figures were far too close to the theory to have arrived there by chance. They had to have been fiddled. But by the time Fisher was writing, Mendel had become a hero – indeed, an icon for the new and somewhat struggling science of genetics, which needed a founder who labored on to find the truth in obscurity and rejection. So what was to be done? Simple. Put the blame on someone else!

> A serious and almost inexplicable discrepancy has appeared . . . in that in one series of results the numbers observed agree excellently with the two to one ratio, which Mendel himself expected, but differ significantly from what should have been expected had his theory been corrected to allow for the small size of his test progenies. To suppose that Mendel recognized this theoretical complication, and adjusted the frequencies supposedly observed to allow for it, would be to contravene the weight of the evidence supplied in detail by his paper as a whole.

Although no explanation can be expected to be satisfactory, it remains a possibility among others that Mendel was deceived by some assistant who knew too well what was expected. This possibility is supported by independent evidence that the data of most, if not all, of the experiments have been falsified so as to agree closely with Mendel's expectations. (164)

One gathers from the last sentence that the more extensive the forgery, the less the likelihood that Mendel committed it.

Reasons for Making Charges

The flip side to the defense of heroes is that sometimes one finds charges of dishonesty being filed in order to expel opponents from the scientific community. Anyone can be wrong, but a cheat is another matter entirely. T. H. Huxley was rather given to this practice, especially when it came to his arch rival, Richard Owen. Owen made claims about the unique nature of the human brain. Huxley did not simply counter him empirically, but argued that one who made such claims had to know that he was lying and should be treated accordingly. In connection with Owen's possible election to the council of the Royal Society, we must ask "whether any body of gentlemen should admit within itself a person who can be shown to have reiterated statements which are false and which he must know to be false." As it happens, although in later years the Huxleyites were to conquer the Royal Society, on this occasion, Huxley's protestations notwithstanding, Owen was elected to the council. Moreover, some felt that Huxley had gone altogether too far in his charges against Owen. The president of the Royal Society, Edward Sabine, thought Huxley's charge "a *very painful* one" and hoped that "we have no occasion to believe that one or other of those in controversy have been 'guilty of wilful and deliberate falsehood'. It would indeed be a painful position for the Society to be obliged to take either side in a *moral* dilemma of so serious a character" (Desmond 1982, 77).

This kind of move continues to the present. Richard Lewontin hates sociobiology, so he not only denigrates Edward O. Wilson's work, suggesting that it is shallow to the point of dishonesty – remember the quote that opens the chapter on humans – but he smears Wilson by association, criticizing him in a book that starts by exposing a genuine fraud, Sir Cyril Burt, who faked results to support his thinking about intelligence testing (Lewontin, Rose, and Kamin 1984). Rather than lingering on this, however, let us turn to the interesting case of Charles Darwin and Alfred Russel Wallace, which shows the issue of integrity working both ways. In June 1858, Darwin received the essay from Wallace, then collecting in the Malay Archipelago – the essay that contained all of the ideas about the struggle for existence and consequent natural selection. After some initial hesitation about what he should do, Darwin put the matter in the hands of his friends Charles Lyell and Joseph Hooker, who arranged for the publication of Wallace's essay along with material by Darwin in the next issue of the *Journal of the Linnaean Society*; and then Darwin sat down and wrote the *Origin of Species*, which was published toward the end of 1859.

Virtually from that moment on, there has been a succession of people who argue that truly it was Wallace who had all of the good ideas, and that Darwin took them away from him and illicitly claimed the credit (Brackman 1980). Darwin, in other words, was a plagiarist. For the record, Darwin was no such thing. The charge is ludicrous. Darwin discovered natural selection back in 1838; he wrote up his ideas first in a 35-page sketch in 1842 and then in a 230-page essay in 1844; and, at the time of the Wallace essay's arrival, was deep into a massive work on evolution and selection (Darwin and Wallace 1958). It is also just plain silly to say that Wallace deserves equal credit with Darwin. It was Darwin who put everything together in a full theory in the *Origin*, and it was Darwin who went on developing these ideas in works like *The Descent of Man*. It is true that in the 1860s, Wallace had very stimulating ideas about selection and that he deserves credit for them, although let us not forget that it was Wallace and not Darwin who got enthused about spiritualism and began arguing that human evolution demands the intervention of non-natural forces.

So ridiculous is the charge of plagiarism – Wallace himself never thought for a moment that Darwin had done ill by him – that one is left to ask, why is it so popular? The reason is not hard to find. Wallace was a far more romantic figure than Darwin – always in pursuit of some wild idea, never able to keep a proper job, and having to live by his pen (or, until he got a state pension, huge amounts of examination paper marking), with a sweet nature and personal sense of right that endears him to those who did not have to live with his foibles (Shermer 2002). He was not a child of the masters of the Industrial Revolution, someone who was never in doubt about his self-worth or financial security, able to beget a large family without having a care or any need to find employment. For many today, Wallace is just a more sympathetic figure than Darwin – a holist and socialist and feminist and antiracist, rather than a reductionist and capitalist and happy supporter of the sexual and racial status quo. Even if one is not going to go the way of spiritualism, Wallace represents a warmer, fuzzier, more directed (especially when it comes to humans) kind of evolution than does Darwin, and so his partisans do all that they can to promote him and his cause.

Things go the other way too. Why did Darwin not simply chuck Wallace's paper into the fire, or sit on it until he had published his own work? Or at the least, get his chums to suppress Wallace until Darwin was firmly in print? In part, the answer has already been given. Darwin was no amateur. He was a professional who had internalized the rules of science. To suppress Wallace, even if it were possible (and it might well have been), would have violated Darwin's own standards – and one supposes those of his friends also. So something had to be done. But here I think we get another pertinent factor. Darwin and his associates were about to push a very radical position, one that was already tainted with infidelity and more. It was part of their strategy of response to show that they were more moral, more respectable, than any other group in Victorian society (Ruse 2005b). They had to show that no one could accuse them of anything but the highest behavior and thoughts. Huxley, although not directly involved in the publication of the Darwin/Wallace papers, was forever going on about science leading to the highest form of moral thought and behavior – and Darwin and Lyell and Hooker endorsed this absolutely. Hence, for Darwin and his friends it was vital that Wallace be treated in a full and generous way. He was not around to make decisions for himself; he asked Darwin to take over the matter of publication, and this Darwin did. But note that Darwin did not do this himself, but got others to do it for him, to give himself the distance and thereby avoid the odor of self-promotion. They were not going to be fools. After some initial worries by Darwin, no one seriously thought that Wallace would get all of the credit. But credit he had to be given, and so he was. A nice solution all around, one that served everyone well.

I have been telling you about Darwin and Wallace as much for the messages to be extracted as for its inherent interest. As we turn now to the broader range of evolution and its history, we find, as expected, other claims of fraud and plagiarism. I do not want to say that all of these claims are worthless. Some of them are very well taken. I do want you to be sensitized to the fact that claims are not always made by disinterested observers of the scene. Sometimes they are made by people who think that the charges can destroy the whole enterprise itself. Give Darwinism the stench of dishonesty, and who needs to worry about its merits or achievements? With this possibility at the back of our minds, let us look at three cases of putative fraud, starting with Ernst Haeckel's dodgy pictures.

Haeckel's Pictures

Jonathan Wells gives embryology a major place in his critique of evolutionary thinking. We learn that it is riddled with fakes, that it is "a classic example of how evidence can be twisted to fit a theory," and that as a result people like Stephen Jay Gould have become accessories to letting fellow evolutionists commit "the academic equivalent of murder" (Wells 2000, 109). Can this really be so? Start with the fact that embryology does indeed have a vital part to play in the evidence for evolution. The fact that the embryos of organisms very different as adults often look very similar as embryos is clear evidence, in the opinion of evolutionists, of common ancestry. Why else would the chicken and the human look alike, except that they came from the same source, and then got torn apart by the forces of evolution? Darwin spoke of it as the strongest evidence for his theory – although he was referring not just to the evidence for the fact of evolution, but also to the selection-driven explanation he felt able to offer. It is only on adults that selection really makes its forces felt, the young being protected by the womb or like situations (in the egg, for instance), and so not only do we not find differences between the young, we should not expect to find differences between the young.

Although Darwin himself worked on the effects of selection, measuring the dimensions of animals when young and when adult, he did not himself dig up the basic facts of embryology. Nor – drawing his ideas from von Baer rather than from the *Naturphilosophen* – was Darwin a major enthusiast for the parallels between embryological development and the history of life. But remember that others, most particularly Ernst Haeckel, a German enthusiast for evolution and self-declared prophet of Darwinism ("Darwinismus," as he called it), did endorse and promote an evolutionary version of the recapitulation thesis – "Ontogeny recapitulates phylogeny," what Haeckel called the biogenetic law (Richards 1987, 1992; Nyhart 1995). Thus in the eyes of Haeckel and a great many others, Darwin's theory became identified with a recapitulatory position. And so it remained, although (increasingly as the nineteenth century went on) more and more exceptions to the biogenetic law were uncovered, and by the middle of the twentieth century the so-called law had been rejected by almost everyone (Gould 1977).

Since by this time Darwinism – natural-selection-driven evolutionism – was the dominant paradigm, there was some predictable inclination to separate it off from embryology, with its rather bad reputation. Such downplaying was reinforced by the fact that the population geneticists omitted development from their modeling. (We discussed this point in Chapter 5.) Hence, embryology went from being the strongest point in the evolutionary picture to being virtually omitted. Things began to change again only two or three decades ago, as evolutionary development (evo-devo) gathered steam and showed its importance and power (Gould 2002). But even now – especially now – this does not mean that recapitulation is back in favor. Not that anyone denies that sometimes one gets (as Darwin argued) sequences that do seem vaguely recapitulatory, and or that anyone denies that at a more general level embryology can be a guide in some way to life's histories.

Mistakes were made, but this is not the same as fraud. Why then should people like Wells – who, given his Moonie background, has strong extrascientific reasons for labeling evolutionists' work as fraudulent – get so excited? The reason goes back to Haeckel and continues with the use that others – right down to the present – have made of him. And here one must admit that there is reason for criticism and for shame on the part of evolutionists. Haeckel was a very gifted artist, and his books are filled with illustrations – diagrams and trees and the like. He illustrated his claims about embryology with drawings of the supposedly similar embryos and supposedly different adults (Haeckel 1874). Unfortunately, he was not entirely scrupulous. Instead of showing actual drawings from life, he used – with minimal (if that) retouching – some of the same illustrations to show embryos of organisms very different. It was hardly surprising that everything seemed so similar! It should be noted that very quickly people picked up on this – there were critics of evolution then as there are today – but instead of remedying the problem, Haeckel bluffed and barracked his way out, arguing that illustrations are always idealized

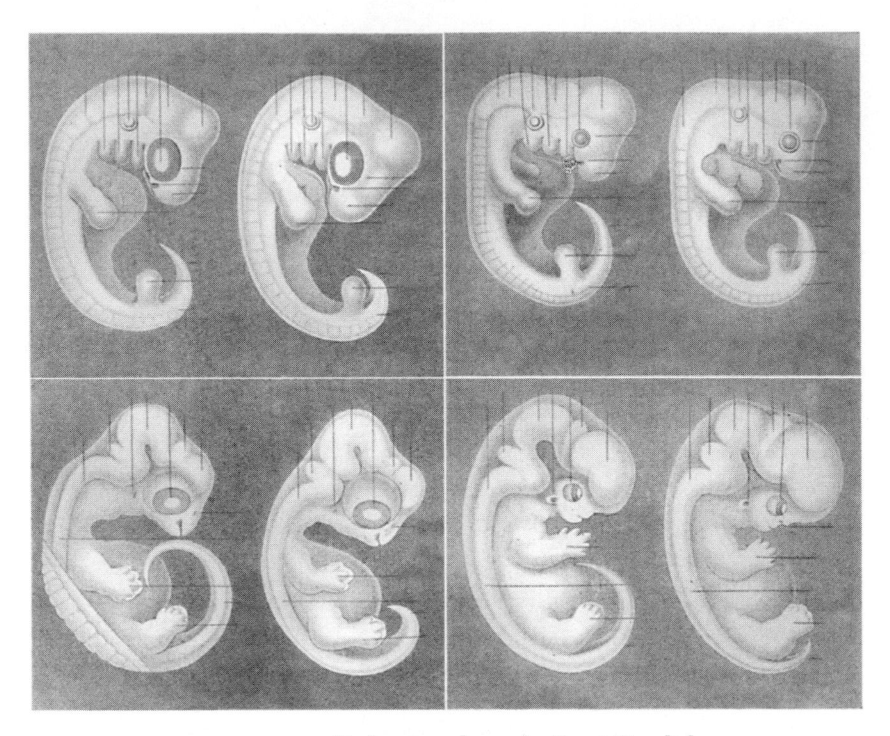

FIGURE 9.1. Embryos as drawn by Ernst Haeckel.

and that he should not be criticized for such a practice. He was (supposedly) merely showing the true facts in a clearer manner (Figure 9.1).

One wishes that Haeckel had been a little more fastidious. He of all people, with his artistic abilities, could have made the same points by using more careful and genuine pictures. He did not, and therein lies the problem compounded. The writers of textbooks tend to share information among themselves – more precisely, they tend to study each other's works and to borrow what seem like good ideas and (apparently) freely available pictures. The Haeckel embryos became the standard fare of biology texts, right up through the 1990s. Over a hundred years after Haeckel's critics had pointed out that his illustrations are a little too good – or, more accurately, not quite good enough – they were still being taken as genuinely representative of embryological development.

Before Wells took up the critical refrain, this bad practice was publicized – by serious embryologists who are committed to the truth of evolution (Richardson et al. 1998). Rightly, textbook writers have shown themselves embarrassed at their shallow practices – shallow rather than deliberately deceitful – and now their books carry either careful and accurate drawings or (more often) detailed photographs. Lazy work – bad work, if you like – was done, but it has been corrected, and whether one wants to say that either fraud or plagiarism was involved (although copying with acknowledgement hardly counts as plagiarism), there is little reason to think that evolutionary studies as such have been irredeemably tainted or destroyed. The pictures are corrected. The work and the teaching can continue.

If Not Dishonest Science, Then Bad Science

Wells does not want the critique to stop here. First, he implies that since Haeckel's pictures were used – since substitutes for Haeckel's pictures are still used – this means that evolutionists are still saddled with the false biogenetic law. To which one can only reply: nonsense. Wells is deliberately conflating von Baer's embryology with Haeckel's embryology (Ruse 1999a). Embryos do look alike. This is the starting point for von Baer as well as for Haeckel. So someone in the von Baer tradition (like Darwin and his successors) can legitimately use the same information as Haeckel (and his successors), even though they reject the conclusions that Haeckel would draw. Ptolemy and Copernicus both drew attention to the motions of the moving stars (the planets), although they came to very different conclusions about the meaning and explanation of the phenomenon. To fault Copernicus for using the data of Ptolemy is simply unfair and scientifically illegitimate. (Nor is it much of a critique to complain, as does Wells, that since von Baer never became an evolutionist, Darwinians should not endorse and use his embryology. Tycho Brahe never became a Copernican, but that does not mean that Kepler was wrong in using Tycho's results.)

The second objection made by Wells – made with the explicit addendum that information is being suppressed deliberately – is that embryos are not that similar. In the early stages of development, from the fertilized egg, organisms are often very different, and it is only after a while that they all converge on the same form (before parting again). One has a kind of hourglass picture of development, and similarity is fleeting. To which one can reply that it is indeed true that the very earliest stages are different, and that this should be acknowledged – although if it has been concealed, one would welcome detailed evidence. Even laypeople know that hens lay eggs and women do not, and every schoolchild taking a biology class learns about what happens after the initial point of conception and the commencement of cell division.

But more important than any of this is the fact that – even though they may differ initially – organisms very different do come together in development. This needs explaining, and evolution is the only viable naturalistic explanation. So, hourglass or not, the pertinent facts remain. And even more significantly, it is now known that if you go back to the earliest stages, there exist considerable parallels or homologies (Wolpert 1991). Physically, initially there are bound to be differences. Reptiles and birds lay eggs, placental mammals do not. Eggs need to carry their own food – yolks – and so the developing organism has to be structured around this large extraneous body in a way that an embryo being fed directly by its mother does not. As soon as one starts to take this into account – the cells of reptiles and birds sit on top of yolks and are thus constrained – the similarities between different pre-embryos start to emerge more and more clearly. And these similarities now are mapped by similarities in genes and in gene action. The same chemical processes seem to underlie what apparently are very different physical bodies.

Third and finally, Wells objects that embryos are really not that similar. One can tell different species apart. This may be, but embryos still are very similar – a lot more similar than adults. The possession of gill slits, for instance, is shared by embryos from fish to humans. Wells objects that gill slits are not really gills, and that therefore there is no evidence here of shared ancestry. But this would be an objection only if one were still working with a biogenetic perspective. Then one would expect the embryos of humans to be like adult fish and have functioning gills. On a von Baer perspective, one expects fish and humans to share features – which they do: gill slits. One can legitimately point out that although gill slits are not gills, they are things very much on the way to becoming gills, so that it is true that embryonic humans do resemble adult fish – but this does not at all concede that they are adult fish. One has something like Haeckel's law, but crucially it is not Haeckel's law.

There is no reason to argue that embryology fails to support the fact of evolution – or that it cannot carry a selection-driven explanation.

Piltdown Man

Sloppy and shabby is one thing. What about cases of outright fraud? Evolution's history is marred by the most famous fraud case in all of science: Piltdown Man (Weiner 1955; Spencer 1990). We have seen that the first real missing link, the discovery of Eugene Dubois, was found in Java. Unmistakably bipedal but with a smaller brain than we humans have, it was named *Pithecanthropus erectus*, although today we put it in the same genus as ourselves, *Homo erectus*. Haeckel seized at once on the significance of Dubois's discovery. In a little book he penned (revealingly entitled *The Last Link: Our Present Knowledge of the Descent of Man*), he saw it as the very piece of evidence long awaited (Bowler 1986). The next major find came in the 1920s, in South Africa (Johanson and Edey 1981). It too was an organism with upright stance, but this one had a far smaller brain. Known informally as Taung baby (because it was a juvenile), it was officially classified in a different genus from humans, *Australopithecus africanus*.

The significance of this discovery was controversial right from the beginning; at first it was considered just an ape and not at all significant for the story of human evolution. It was not until the 1940s that *Australopithecus* was recognized for the significant finding that it really was. Why the delay? One major factor was that the animal came from Africa, and (particularly in the light of Java man) most informed people were looking to Asia. Taung baby just did not fit in, and anyone who knows anything at all about science will realize that prior convictions and expectations are a significant factor in molding scientific opinion. Also, a lot of people simply did not like the smooth upward rise that Haeckel showed and that Taung baby would seem to confirm. There was agreement, of course, that we had evolved and that ultimately we had evolved from beings with small brains. But there was a desire to push this back as far as possible. People did not want to be too closely associated with the apes – even the Neanderthals were now out of favor and portrayed as highly brute-like and not all respectably human. Indeed, white people (who were, after all, the scientific leaders in paleoanthropology) did not want to be too closely associated with their fellow darker humans and wanted many years of evolution independent from other groups.

Fortunately, there was what was thought to be good evidence for this assumption of long-time separation: Piltdown Man. In southern England, around 1912 (the exact date of first discovery is clouded in mist), an amateur archaeologist, Charles Dawson, unearthed pieces of skull and jaw which seemed to confirm that precisely the required sorts of humans had lived and thrived, long before the present. These were humans with massive brains – virtually as big as ours, in fact – and yet clearly

FIGURE 9.2. Piltdown Man.

primitive in other respects, particularly in the lower face and jaw. Conferring authenticity, Arthur Smith Woodward, a keeper at the British Museum (Natural History), became involved in the discoveries, as well as the young French priest/paleontologist Pierre Teilhard de Chardin. Quelling doubters, a year or two later some really major pieces of evidence came to light. (Supposedly Dawson found these new fossils in 1915, although they were not announced by Woodward until after Dawson's death in 1916.)

The Piltdown hoax, as it is now better known, was revealed as a fraud in the early 1950s. It is hardly surprising that Piltdown Man had a brain as big as ours. The key skull was a human skull! Nor was it surprising that the lower face was primitive and ape-like – the jaw and teeth that were recovered came from an orangutan. The pieces were suitably shaped and stained, and then the awkward bits (precisely those bits that would cast doubt on the brain and jaw being from the same animal) were broken off and thrown away (Figure 9.2). The remarkable thing about Piltdown Man was not that the fraud was eventually uncovered, but rather that it lasted as long as it did. It was a crude job. As soon as anyone looked, you could see all sorts of file marks and such things, including evidence of staining rather than weathering through time. There is one piece that actually looks like a cricket bat, as though the trickster were really trying to make people look stupid. Even without physicochemical methods of dating materials, the hoax ought to have been spotted early on. Indeed, to their credit, some people did always feel that it was highly and uncomfortably anomalous. But it fit precisely what most people were after – almost too patly, one might say (especially when more relevant bits appeared almost to order) – and there are none so blind as those determined to see. People were nothing if not this, especially English people, who were highly sensitive to the proud place that England now possessed in the search for human ancestors. The Germans might have those nasty Neanderthals, but fair Albion has been home to the greatest prize of all.

Culprits?

The possible identity of the perpetrator of Piltdown has filled more books than has the quest for the identity of Jack the Ripper – with about as much success. On the internet, in ten minutes I found more information on the topic than I truly need for one lifetime. To be honest, the identity does not really matter, which is part of the attraction. Some of the suggested suspects rather boggle the imagination – although, unlike the Ripper, no one yet has suggested that the Piltdown hoaxer was a son of the prince of Wales. (The hoax may not have been a great work of art, but it required more energy and gumption than one generally associates with British royalty in this or the last century.) One far-out suggestion is Sir Arthur Conan Doyle, the author of the Sherlock Holmes stories. He was a keen spiritualist and had a keen dislike of scientists, who regarded his enthusiasms with contempt. Hoaxing them all like this would have been very satisfying.

The most recent candidate is one Martin Hinton, a curator at the British Museum. He has been indicted on grounds of bits and pieces of supposedly incriminating evidence found in his effects discovered after his death (Highfield 1996; Gee 1996). But it appears that he cannot have been the sole perpetrator – he was simply not around at some of the required times – and the evidence may not be what it seemed. (Particularly suggestive was a discovery in Hinton's effects of various chemicals that were needed for "aging" the orangutan jaw; but Hinton's chemicals do not match exactly the chemicals used on Piltdown.) General suspicion has always centered on Dawson, who had a bit of a reputation for being shifty, and probably this is not far off the mark. Woodward may well have been a dupe – it is interesting to note that his specialty was fish rather than humans.

The story continues, with no prospect of an end. What moral are we to draw from all of this? It was undoubtedly a case of fraud, although the motive must remain a mystery, given that identity of the the perpetrator(s) is a mystery also. One doubts that it was something done deliberately to derail the course of evolutionary studies. One suspects that it was done either to make a fool of one of the participants or simply as a jape that got out of hand. Who would dare confess after it got taken so seriously? Perhaps it did hold back the course of human evolutionary studies for a while, although probably these would have been slow to progress anyway, until more fossil evidence was unturned in Africa and there was a stronger sense that Darwinian selection must be the primary cause of evolutionary change, including human evolutionary change. One has no sense that the hoax held back the course of evolutionary studies generally. Work on genetics continued apace in the 1910s, and the population geneticists set to work without encumbrance in the next decade. Given the way in which so many so enthusiastically took up the Piltdown findings, one does have confirmation of the extent to which some people – especially those focusing on human origins – let their hopes and prejudices shape their conclusions. It would probably be naïve to think that this is a practice that is entirely absent today.

What is interesting – although perhaps predictable – is the extent to which the Piltdown hoax is still used as a weapon in battles over evolution. Jonathan Wells makes very sure that it features prominently in his attack on evolutionism. It is given a role by believers as well. Around 1980, Stephen Jay Gould decided that thoughts of biological progress are ethically pernicious – they suggest that Jews and others are inferior to Anglo-Saxons – and should be opposed. So, as part of his campaign against progress, Gould took aim at the most enthusiastic biological progressionist of the twentieth century – Father Pierre Teilhard de Chardin. Gould (1980b) went about this by arguing that Teilhard (who was present at some of the relevant digs) was the person responsible for the Piltdown hoax. The evidence for this charge was thin to the point of nonexistence, but Gould knew what he was about. In the spirit of Huxley before him, by labeling Teilhard as a hoaxer – as the perpetrator of a major fraud – Gould could argue that Teilhard is no true scientist and hence has no call on our beliefs. Teilhard might argue that biology is progressive, but what standing has he to tell us what to believe? His career as a paleontologist counts for naught. As I have said, not everyone regards fraud in science as an unmitigated evil. Even the darkest clouds can have silver linings.

Peppered Moths

Let us conclude with the peppered moths, the most famous example of evolution in human lifetimes (Majerus 1998). We know that, from the start, it was pegged as a case of natural selection in action. And so it has remained as the paradigm case of natural selection powering evolution. Although some challenged the thinking, it was reaffirmed and practically canonized in the 1950s, thanks to the work of H. B. (Bernard) D. Kettlewell (1973), who performed many experiments and studies on the moths, concluding that differential predation is indeed the chief causal factor. Kettlewell also pointed out (in the footsteps of J. W. Tutt) that as and when pollution practices are changed and reversed, one might expect to see the ratios of peppered to dark (melanic) forms being reversed, with the possibility that all could revert to the situation before the Industrial Revolution got under way. And in fact this has happened; indeed today in parts of Britain (and elsewhere in the world) the melanic form seems on its way to elimination and extinction.

Why then has this example of natural selection in action recently come under very heavy fire? Critics argue that Kettlewell's work was flawed – trimmed and shaped to make the desired conclusions – that alternative explanations were dismissed out of hand, and even worse, that there was outright fraud or fakery, especially in the pictures used to illustrate the phenomenon (Wells 2000; Hooper 2002). More specifically, Kettlewell performed experiments that had little to do with the natural conditions of the moths; he deliberately suppressed crucial pieces of evidence, especially about the normal resting places of the moths; and the pictures that everyone – especially textbook writers – use are (knowingly) fraudulent, using dead moths glued to trees in what are known to be unnatural

FIGURE 9.3. Peppered moths against tree backgrounds. Photos courtesy of Brent P. Kent. From J. Hooper, *An Evolutionary Tale of Moths and Men* (New York: Norton, 2002).

positions (Figure 9.3). The changes in pollution and the dirty trees and atmosphere do not track the changes in melanic-to-normal proportions, either geographically or temporally. The wrong ratios are found in wrong parts of the country, and changes occur over time, in pollution and in ratios, that too often do not track each other. And, even worse, today we know that the whole story is a mess, simply because birds do not see things in the same way as humans – they are sensitive to ultraviolet light – and hence what may seem black to us can seem white to them, and what may seem white to us can seem black to them. In short, as one eminent evolutionist has said, the whole situation is rather like finding out that there is no Santa Claus, and that all the presents come from your parents (Coyne 1998).

Digging out from beneath these charges will be no easy business, but we can try. Let us agree that judged by today's standards (hardly a surprise), Kettlewell's work of fifty years ago leaves much to be desired. Let us agree also that the story of industrial melanism is (hardly a surprise) much more complex than anyone would have thought just a few years ago. Let us also agree (perhaps a nasty surprise) that the photos used to illustrate industrial melanism are much more artificial – call them faked, if you will, although we might see reason to question this – than most readers know (and, one suspects, than most authors know). What is left? Well, start with Kettlewell. One of the big issues that he was facing – perhaps the most basic issue – was whether birds do indeed eat the moths. This he established unambiguously, indeed getting the ethologist Niko Tinbergen to go along with him to wild areas and actually photograph birds in the act of catching and eating moths. He was also able to show that there is good evidence, both experimental and natural, that there is differential predation. Black forms get eaten more often in rural areas and light forms in industrial areas.

> Kettlewell's famous mark-release-capture experiments in Birmingham and Dorset provide strong evidence for the differences in fitness of the forms in polluted and unpolluted environments. Furthermore, his observation that live moths, which had been released on to tree trunks, were differentially preyed upon by birds provides evidence that the intensity of bird predation on the morphs varies according to habitat. (Majerus 1998, 126)

But are there not a number of nasty counterfacts? What about the resting places of the moths? It certainly does seem true that these moths do not generally rest on the upright trunks of trees. This does not mean that they do not rest on trees, but rather not on the visible upright trunks. They are more likely to rest on branches or just below them or under them. This is going to give them more protection. This means, however, not that color camouflage is now unimportant, but rather that it is somewhat more complex. The same is true of bird vision (Howlett and Majerus 1987). If anything, in fact, the new findings underline the strength of the selective hypothesis. Different lichens grow on different sides of branches, and it seems that although we might not find the moth-color differences particularly effective against the different vegetable bases, from the moths' viewpoint the birds do.

One area where there does seem significant reason to modify the Kettlewell work concerns migration. He did not take this to be a significant factor. It seems that the moths can and do travel quite significant distances. This obviously could mess up correlations between backgrounds and colors. But, if anything, this seems to count in favor of the adaptive-melanism-against-predation hypothesis! Take East Anglia (Norfolk and Suffolk) – a rural part of England if any is. There seems to be a higher proportion of melanic forms there than theory would lead one to suspect. However, what one must not ignore is the prevailing winds in Britain from west to east – in particular, the winds from the industrial Midlands – Birmingham, Coventry, Wolverhampton – that blow across to East Anglia. I remember very vividly, in the Black Country town of Walsall, the near-gales as one biked to and from school in the early 1950s. (This part of the Midlands – Walsall is ten miles from Birmingham – is so called because of the pollution from heavy industry.)

Taking these weather conditions into account leads one to expect that in the East there will be an elevated number of melanics. Taking into account similar findings elsewhere in the world, and like patterns in other species (in and beyond the Lepidoptera), the leading expert concludes:

> My view of the rise and fall of the melanic peppered moth is that differential bird predation in more or less polluted regions, together with migration, are primarily responsible, almost to the exclusion of other factors. I do not support the view that non-visual selection plays a very significant part in Britain. Lack of correlations and concordance between simulations and observed frequencies are, to my mind, more likely to be due to lack of understanding of the resting behaviour of the moths and their ecological interactions with their predators, than selective factors whose nature has yet to be identified. (Majerus 1998, 155)

But what about the charge of outright fraud? What about the pictures? They are certainly not what they seem. You see moths on the trunks of trees, and you think you are seeing the real thing. You are not. Apparently they really are dead moths, glued on. As such, this does not make the situation fraudulent. Even though the moths are dead and even though the moths generally do not rest on the upright trunks, it is legitimate in an experiment, as a first-order probe, to simplify the situation this way. You want to see if birds do take the moths, you want to see if the moths do seem to stand out differentially, and you want to see if the birds catch their prey differentially. It is much easier to manipulate things this way than to use live moths (that may or may not have been released), which are up trees in awkward places – awkward because that is where they are going to be to escape their predators.

All models, all experiments, simplify, so the mere act of simplification is not a fault. What would be wrong would be to take this as the final experiment, and to conclude that now all answers have been given. But no one pretends that this is what has been done. It has been the very experimenters themselves who have pointed to the artificiality of their work, and who have thus been led to try to introduce more realistic – albeit more complex – conditions. One could, for instance, put dead moths above and below branches with and without lichen, and see how color factors influence predation. More realistic, even if still not completely realistic. This is how science works. To think otherwise and to criticize on theoretical grounds is to misunderstand the role and nature of models and experiments.

What is unfortunate, to the point of being fraudulent, is using the pictures, particularly as pedagogic aids, without some qualification explaining the artificiality of the situation. Not the act of using the pictures themselves – I see no reason why they should not be used in textbooks. In fact, with the kind of qualification given in the last paragraph, one could both demonstrate the striking phenomenon of industrial melanism (however caused) and use the opportunity to give a little lecture on the nature of experimentation – how one refines one's models to get ever-more-accurate simulations. But using the pictures without qualification does give a distorted picture of the ease of confirmation of the melanism hypothesis. A superb edition of *Scientific American* devoted to the topic of evolution carried an article on adaptation by Richard Lewontin (1978) – an article that began with pictures of the moths in living color. Probably Lewontin himself did not find and place these pictures, but the impression given is certainly that we are looking at real-life phenomena – phenomena that seem to prove more than they really do. This is wrong. Either someone did not do their homework and find out that the pictures were artificial – sloppy if not intentionally dishonest – or someone did do their homework and suppressed the evidence – dishonest (or mistaken and condescending about what the reader should know). All of this is a pity, especially since doing things the right way has demonstrated that the peppered moths really do earn their jeweled place in the crown of Darwinism.

Conclusion

What do we conclude overall? Three examples cannot make a definitive case one way or another, but they are three of the most commonly cited examples of the fraudulent nature of evolutionary studies. Three points can be made. First, evolutionary studies have had cases of dishonesty, of various grades. Perhaps one judges less-than-candid pictures in textbooks less harshly than the worst kind of transgression, but they are to be regretted nevertheless. The Piltdown case, done for whatever reason, turned out to be one of the major scientific frauds of all time. Second, it is hard to say whether evolutionary studies is worse in this respect than other fields, or whether fraud is endemic in evolutionary studies judged just in its own right. The nature of the beast is that a really successful fraud gets away with it and no one notices. There is a self-limiting factor here. The more important the piece of science at stake, the more likely fraud (or plagiarism) is to be uncovered. Judged in this way, one does not have the impression that evolutionary studies is significantly more plagued by or prone to fraud than other areas of science, nor indeed that fraud is that common in evolutionary studies judged on its own.

The third and most important point is that it is just plain silly to claim that fraud in evolutionary studies corrupts all that is done in the field. The Piltdown hoax probably set back paleoanthropology for a while, although, as pointed out, it is hard to make definitive judgments about this field. Haeckel's own work was probably not the most profitable, but if one criticizes the enthusiasm for the biogenetic law (and one should), remember that there is a difference between being fraudulent and just being wrong. The question is whether Haeckel's shifty pictures actually retarded the course of science, and this is not so very likely. The same is true of the melanic moth case. Of course, things have moved on from Kettlewell, but this is the nature of science. One does not have to be a follower of Karl Popper to agree with him that the course of science is one of slowly and sometimes painfully groping gradually toward the truth – which we may never achieve. Fifty years after Kettlewell set to work, we know a lot more – and with the new powerful techniques (like DNA fingerprinting) we are set to learn even more. Far from being dishonest science or even bad science, this is vigorous science. This is science at its best. This is science making full use of Darwin's legacy.

CHAPTER TEN

Philosophy

Darwin's theory has no more relevance for philosophy than any other hypothesis in natural science.

L. Wittgenstein (1923)

Many Anglo-Saxon philosophers would judge the Austrian-born thinker Ludwig Wittgenstein to have been the most important philosopher of the twentieth century. With respect to evolution, writing at the time of the First World War, he set the pattern for a whole generation of philosophers – although, in respects, Wittgenstein was only repeating what was commonly assumed by the fraternity. The American Pragmatists had been very interested in evolution – although the greatest, C. S. Peirce, never much cared for Darwinism – but the twentieth century opened on a low point. In 1903, in his *Principia Ethica*, the Cambridge philosopher G. E. Moore trashed Spencer's thinking about ethics, and for many this was definitive. There were those who took evolution seriously. Henri Bergson was one, although we know where he stood on naturalism and hence on Darwinism. Later there were, in Britain and America particularly, others who wrestled with evolutionary ideas. A. N. Whitehead (1967), who taught at Harvard in the latter part of his life, comes first to mind – although, like Bergson, he had quasi-religious axes to grind. Generally, respectable Anglo-Saxon philosophers stayed away from evolution, and Darwinism was a dirty word – or really, not much of a word at all.

Did things change later in the century, after the fusion of Darwinian selection with Mendelian genetics, when biology finally had a proper and professional evolutionary theory? Not as one might have wanted.

Notoriously, Wittgenstein and Karl Popper were not close friends. But when it came to evolution, they were so far apart that they came back together. Far from thinking Darwinism irrelevant to philosophy, Popper thought it *was* philosophy! "I have come to the conclusion that Darwinism is not a testable scientific theory, but a *metaphysical research programme* – a possible framework for testable scientific theories" (Popper 1974, 168). To his credit, Popper tried, he really tried, to take Darwinism seriously as a science, and perhaps toward the end of his long life he succeeded. But one senses that there was always ambivalence. Somehow, Darwinism did not measure up as a proper science – one that was genuinely falsifiable (the Popperian criterion of demarcation) – at least, not by the standards of physics and chemistry, the standards by which all else must be judged. The result was happy neither for science nor for philosophy. On the one hand, Popper could not appreciate properly one of the triumphs of modern science. On the other hand, Popper created a bastardized form of Darwinism that corrupted his philosophy. At least Popper knew that Darwinism was important – which may or may not have done more for his philosophy than it did for Wittgenstein.

My aim in this chapter is to defend Darwinism from false (or misguided) friends as well as from real enemies – and, worst of all, from the indifferent. If we humans are an end product of a long, slow, law-governed process of natural selection rather than the favored of God created miraculously on the Sixth Day, Darwinism simply has to be relevant to philosophy.

Traditional Evolutionary Epistemology

The critics have had good reason to be critical. Some dreadful stuff has been fobbed off under the umbrella of evolution, and even when it is not that dreadful, some very shaky assumptions have been incorporated (Ruse 1998). We are already prepared for this, because in our discussion of Darwinism as a religion we saw that many (starting with Herbert Spencer and ending with Edward O. Wilson) have thought that evolution of some form or another provides both the dictates of morality (what is known technically as substantive or normative ethics) and its foundations (what is known technically as metaethics). Thus, for instance, Wilson thinks that evolutionary biology forces on us the obligation to preserve biodiversity, and he also thinks that this is justified by the fact that evolution has made humans biophilic – that is, with a love and need of the world of animals and plants (Wilson 1984). But while it may indeed be true that we should preserve the Brazilian rain forests – I, for one, think we should – and while it may indeed be true that a love of nature is ingrained in our genotype – as a happy city dweller I am a little less sure about this – it is clear that there is something wrong here. Moore expressed his worry by saying that at this point evolutionists are committing what he called the "naturalistic fallacy" – which is a version of an older critique of all such arguments first made by the Scottish philosopher David Hume (1978). He objected to any move that goes straight from a statement of fact to a statement of obligation – in Wilson's case, from the claim that we have a need of biodiversity to the claim that we ought to preserve biodiversity. Hume's point is that these are different things, chalk and cheese, and that the one does not imply the other.

My experience is that people like Wilson tend to be supremely indifferent to charges like these. They point out that in science one is always going from talk of one kind to talk of another kind – from talk about little balls bouncing around in a container to talk about the pressure and volume of gas – and they see no reason not to do it here. But as with the arguments, we have already at hand the real reason to fault their thinking. Wilson and traditional evolutionary ethicists before him, back through Julian Huxley to Herbert Spencer, justified their thinking by appeal to biological progress. They thought that the evolutionary process was not random, going nowhere, but reaching ever upward until it arrives at our species, *Homo sapiens*. For the traditional thinker, the moral directives come from the fact that unless we follow them, progress will come to an end – decline may even set in – and the justification comes from the inherent worth of progress itself. Humans are higher and better than others, and that is reason enough to cherish us and to want us to succeed in the future. Unless we preserve the rain forests, humans will die out, and that is a bad thing.

But we have seen reason to think that the progress of evolution is problematical, to say the least – it is certainly not a given if you are a Darwinian evolutionist. Which means immediately that, however worthy the substantive ethical directives of the traditional approach may be, they have no true foundation in evolution. Hume was right. You cannot justify a moral directive with a factual claim, not even in the case of evolution. A point, incidentally, that T. H. Huxley – someone who had actually written a book on Hume – realized and argued with vigor. In the early 1890s, in his last great essay, "Evolution and Ethics," Huxley pointed out that evolution leads to natural weapons of violence and destruction – the claws and ferocity of the tiger, for instance – and that it is hardly the case that morality demands that we support such attributes. On many occasions, that which is right involves fighting that which has evolved rather than supporting and promoting it.

The genuine Darwinian need not fear the criticisms of traditional evolutionary ethics. They are well taken, but it is not his or her theory that is standing behind the philosophy. Very similar remarks can be made in the case of traditional evolutionary epistemology – the attempt to explain and justify knowledge by an appeal to evolution (Campbell 1974). Interestingly here, T. H. Huxley was one of its first exponents and boosters, for he picked up right away on how one might go from biological evolution to some kind of cultural evolution, showing the development of science and its ideas. In the biological world, we have a struggle for existence between organisms, and the consequent selection of the fitter. Likewise, in the scientific world we have a struggle for existence between ideas and the consequent selection of the fitter. Copernicus beats out Ptolemy because at some level he has a fitter theory. In post-Huxley examples, Einstein beats out Newton, and molecular biology beats out Mendelian.

Down through the years, many have toyed with ideas like these. A recent exponent was Popper (1974), who thought that the process of trying to falsify theories taken over time leads to a kind of evolutionary process, as one gets ever closer to the truth about objective reality. He pictured the course of science as follows.

$$P_1 \rightarrow TS \rightarrow RR \rightarrow P_2$$

You start with a problem, then you offer a tentative solution; you put it to test, trying rigorously to refute it, and then in the process another problem emerges. What is interesting is where the initial problem leads to rival suggestions for solution. They have to battle it out to win acceptance by scientists – better supporting evidence or whatever – and then this new winning solution in turn throws up more problems to be solved.

This is all fairly general, but probably not a bad picture of the course of science. It is evolutionary of a kind also. But is it very Darwinian? Is the analogy close enough to throw real light on the development of science and in turn to bolster our confidence in evolution through natural selection? This is a different question, and the prospects are not promising.

Most significantly, as critics have long pointed out, there is a fundamental difference between the new variations of biology and the new variations – or tentative solutions – of science. As Darwin always insisted, the former are random, not appearing according to need – all of the creative work is done by natural selection. In the case of science, however, the new variations are rarely (if ever) random – they come after considerable struggle and labor, and appear according to need. Darwin himself provides the paradigm case. Coming to natural selection was no chance event. He knew that it was the top prize in his group – the mystery of mysteries – and after becoming an evolutionist in the spring of 1837, he worked long and hard to find the causal answer, eventually hitting on it eighteen months later, at the end of September 1838.

Prima facie, this is a major disanalogy. Some – for instance, the philosopher David Hull (1988) – think that this simply points to the fact that all analogies have some differences and that we should not pick this out as something special or particularly problematic. (See also Toulmin 1967; Richards 1987.) However, some differences are more significant than others, and this – especially in light of Darwin's repeated and adamant insistence that variations not be directed – seems about as significant as something can get. It combines with other factors. In science, unification is absolutely crucial. The really big steps forward – Darwin's theory of evolution through natural selection – come from consilience. A number of different areas are unified under one sweeping hypothesis. The biological equivalent is hybridization. This is common in plants, although always occurring between relatively close relatives and not between absolute strangers – the mark of the really powerful consilience. Traditional physics always kept the Earth and the heavens apart, and yet Newton's genius was to show that Kepler and Galileo were talking about the same thing. It is true that one had the unification of prokaryotes necessary to make eukaryotes, and that there is always some lateral gene transfer, but the one instance is unique, and the other seems not to involve major structural-gene transfer, so again there seems not to be a real analogy with science.

There is a great difference between biological change and scientific change. The former is random or undirected – there may be a rise in complexity or some such thing, but ultimately there is no overall direction. In science, we think that there is such direction. Copernicus was better than Ptolemy, Newton was better than Aristotle, Darwin was better than the Creationists, Mendel was better than Darwin. We may never get to the ultimate truth, but we are closer to it than we were a hundred years ago, and a lot closer to it than we were two and a half thousand years ago. The same is just not true of biological evolution. Admittedly, some (like Thomas Kuhn) want to deny that there is any real progress in science – for that reason, Kuhn (1962) rather liked an evolutionary analogy – but this is surely not true. The cumulative effect of the earlier chapters has been to show that only by jettisoning almost every aspect of human rationality and every standard of evidence can we genuinely say that the pre-Darwinian picture of origins (which was hardly a picture at all – certainly hardly a scientific picture) is equal or superior to a picture of evolutionary origins for organic beings. There really is progress!

In other words, as with ethics, traditional evolutionary epistemology falls afoul of the fact that Darwinian theory gives no reason to see the picture of life's history as genuinely progressive. There is a crucial disanalogy. Scientific change is really not the same as cultural change, and again, those who criticize traditional evolutionary epistemology leave the Darwinian picture of life untouched.

Darwinian Epistemology

Is there a better and truer way of bringing Darwinism to bear on philosophy? Richard Dawkins (1976) would have us consider culture as being in some sense on a par with biology – whereas in biology the secret is the transmission of the biological units of information, the genes, in culture the secret is the transmission of the cultural units of information, the memes (Blackmore 2000). However, thus far this approach seems not to have achieved much more than a fancy redescription of the phenomena. No interesting new predictions or anything like that. You could, I suppose, think of Darwin's theory as being a meme and a pre-Darwinian biblical account as being another meme, and think of them wrestling for supremacy in someone's mind, but if this says more than is said by traditional evolutionary epistemology (with its difficulties), it is hard to see how.

So we ask again: is there a better and truer way of bringing Darwinism to bear on philosophy? Darwin himself certainly thought so. In the case of epistemology (to start with that), he saw that the key must be taking a literal approach rather than (as above) one of analogy. We must somehow see knowledge as embedded in our biology. That is to say, we must see knowledge as adaptive – biologically adaptive. Obviously, we cannot just state matters crudely and leave things at that. Science is part of culture – we have seen that – and no one, especially not a Darwinian, is going to deny that in some respects humans escape their biology through culture. In science, it is just plain silly to pretend that every last belief has direct biological value. No one is surprised that Mendel, who was right about heredity, had no children, and that Darwin, who was wrong about heredity, had ten children (seven surviving to adulthood). So whatever else, we cannot simply hang our thoughts and actions on natural selection and leave it at that.

More plausibly, we look to the underlying principles that help us to build up science, and see if they can be rooted in our biology. What are we referring to here? At a basic level, I presume we need logic and mathematics. Without these, no science, especially no sophisticated science, is possible. But we need more than this, for the point of science is interaction with and understanding of the world of empirical experience. We start with problems, be these theoretical (like trying to understand why the Galapagos finches have the peculiar distribution that they do) or practical (like trying to build a bomb that will end World War Two at a stroke). We need some way of structuring experience, using logic and mathematics as indispensable tools. And this brings us back to the epistemic values. Science is informed and structured and built by and through rules that reflect epistemic desirabilities – coherence, consistency, unificatory power, predictive ability and fertility, simplicity.

It is these values – together with the principles of logical and mathematical reasoning – that must be related to adaptation and natural selection (Butterworth 1999; Pinker 2002). In other words, in a move that is now obvious, the Darwinian assumes simply that the rules of mathematics and logic, the basic beliefs about causality and the like, the epistemic values or principles, are not simply cultural ephemera that were invented by people (either at the time of the Greeks or two thousand years later during the Scientific Revolution) but are at some level ingrained in our biology (Ruse 1998). They are part of our genetic heritage just as much as our physical features and our emotional inclinations like sexual desires. One thinks mathematically because one is biologically disposed to do so, and one is attracted to simple and elegant theories for the same reason. And why should this be so? Very straightforwardly, because those of our would-be ancestors who thought mathematically and logically and preferred the simple to the complex tended to survive and reproduce, and those that did not, did not. The great twentieth-century American philosopher W. V. O. Quine (1969), writing about causality and why we think the future is going to follow the same rules as the past (the problem of induction), put his finger right on the problem and the solution:

> One part of the problem of induction, the part that asks why there should be regularities in nature at all, can, I think, be dismissed. *That* there are or have been regularities, for whatever reason, is an established fact of science; and we cannot ask better than that. *Why* there have been regularities is an obscure question, for it is hard to see what would count as an answer. What does make clear sense is this other part of the problem of induction: why does our innate subjective spacing of qualities accord so well with the functionally relevant groupings in nature as to make our inductions tend to come out right? Why should our subjective spacing of qualities have a special purchase on nature and a lien on the future?
>
> There is some encouragement in Darwin. If people's innate spacing of qualities is a gene-linked trait, then the spacing that has made for the most successful inductions will have tended to predominate through natural selection. Creatures inveterately wrong in their inductions have a pathetic but praise-worthy tendency to die before reproducing their kind. (p. 162)

One is not arguing now for innate knowledge in the sense that John Locke considered and dismissed in *An Essay Concerning Human Understanding*, but more for capacities that lead one to think in various ways (something that Locke considered and accepted). The claim that we have these innate principles (as the capacities manifest themselves in development) – or, as they have sometimes been called, "epigenetic rules" (Lumsden and Wilson 1981) – is an empirical one, as is the claim that we have these principles because they are adaptively advantageous. At the rough-and-ready level, there is much plausibility in these claims, and this is surely the reason why many biologists and philosophers have been attracted to the position. Darwin himself set the pattern and saw the way: "Plato says in Phaedo . . . that our 'imaginary ideas' emerge from the pre-existence of the soul, are not derivable from experience. – read monkeys for pre-existence" (Barrett et al. 1987). The fact is that an intelligent primate that could see that three oranges were more than two oranges, or that thought a cave where two tigers were seen to enter and only one emerged was a place of danger, or that simplicity is a virtue when a lion approaches, was ahead of a primate that thought and did none of these things. The proto-human who saw beaten down grass and bloodstains and heard growls and who yet said: "Tigers, just a theory not a fact," was less biologically fit than the proto-human who started running and who is still at it.

This is just a sketch. Evolutionary psychologists have pointed out that there is probably more to the story than this. Much human reasoning will have developed in social situations, and we should therefore expect that we will spot correct logical inferences more easily in such situations rather than in abstract problems (Cosmides 1989; Barkow, Cosmides, and Tooby 1991). Well-known psychological experiments suggest that this is true. None of this threatens the point just made – rules of reasoning are part of our biology – showing rather why so many have trouble reasoning logically when in unfamiliar or abstract territory.

Circularity?

We can now complete the picture sketched earlier of the nature of science. I have argued that science – Darwinian evolutionary theory in particular – is a blend of the objective and the subjective, the transcendent and the cultural, discovered and created. As it tries to make sense of experience, it achieves objectivity through its pursuit of epistemic excellence. It is subjective, cultural, through the metaphors that are an essential tool toward understanding and satisfaction of the epistemic. But, for the Darwinian epistemologist, the epistemic values themselves are not reflections of some absolute reality – forms in a Platonic heaven or whatever. Rather, they have the pragmatic origin of having proven themselves in the struggle for existence and having thus been selected to inform and structure our thinking. The foundations of science, therefore, are a subtle blend of the biological and the cultural – epistemic and metaphorical coming together to make one whole.

Are people like me being a little too casual at this point? Although I am hardly a full-blooded subjectivist, a "social constructivist," have I nevertheless given away too much? Could it be that, having let in the virus of subjectivity, we are going to find it impossible to keep it from infecting everything that we do and think? Most particularly, isn't something like Darwinian evolutionary theory particularly susceptible to this disease? This is the fear – more precisely, the happy and somewhat triumphant conclusion – of Alvin Plantinga (1991, 1994), the already-encountered chief philosophical scourge of Darwinism today. As it happens, given that he is an ardent Christian, a Calvinist, Plantinga thinks that he can appeal to God to avoid the problems of subjectivity. But whatever the strength of his Christian alternative, he believes that the charge of unacceptable subjectivity stands alone.

Plantinga's argument is simple and (if well taken) powerful. If we are Darwinian evolutionists, then Wittgenstein could not be more wrong. Darwinism is important for our powers of thinking and action, and it reflects into the products of our efforts. But consider for a moment the old joke about Darwinism – it values only the four effs: fighting, feeding, fleeing, and reproduction. There is nothing here about getting our thinking into right order with reality, nothing about gaining (in the language of Karl Popper) "objective knowledge." What works, what succeeds, is what wins. Interestingly, notes Plantinga, Darwin himself realized this, writing: "With me the horrid doubt always arises whether the convictions of man's mind, which have been developed from the mind of the lower animals, are of any value or are at all trustworthy. Would anyone trust in the convictions of a monkey's mind, if there are any convictions in such a mind?" (Plantinga 1993, 219, quoting Darwin 1887, 1, 315–16). Cleverly, Plantinga refers to this as "Darwin's doubt," although perhaps he is being a little too clever here; having expressed this doubt, Darwin at once excused himself on the grounds that he was moving into murky philosophical waters he did not feel competent to navigate. Admiral Plantinga has no such hesitations. Darwinism as a theory is self-refuting. It is all a bit like one of those language paradoxes: "This statement is false." Is the statement itself true or false? If it is true, then it must be false, and if it is false, then it must be true. If Darwinism is true, then there is no reason to think that it is true, and probably good reason to think that it is not true. If Darwinism is false, then it might be true, and that is an even less helpful conclusion. Either way, drop Darwinism!

As we have seen already, part of the trouble when dealing with Plantinga is that he takes so cavalier an attitude toward the things that he criticizes, Darwinian evolutionary theory particularly. Before we accept his conclusion, we should at least look at the reasons why we might think that Darwinism is going to deceive us – why we might think that Darwinism is going to deceive us systematically, leaving us no recourse to find and correct the mistakes. Plantinga himself does not give us much help here. Drawing attention to a posh dinner at an Oxford college, where the philosopher A. J. Ayer (author of the notorious *Language, Truth, and Logic*) debated issues of religious belief with the biologist Richard Dawkins, Plantinga suggests that from a Darwinian perspective all of this could be a charade, a kind of mental picture show with no connection to reality. "It *could* be that one of these creatures [Ayer or Dawkins] believes that he is at that elegant, bibulous Oxford dinner, when in fact he is slogging his way through some primeval swamp, desperately fighting off hungry crocodiles" (Plantinga 1993, 224).

Well, actually, with all due respect, it could not be. At least, not on a Darwinian perspective. Fighting crocodiles is not at all aided by thinking that you are somewhere else, and certainly not by thinking that you are eating and drinking too much in a medieval university hall. It is true that Darwinism says that it is reproduction that counts, but by and large Darwinism expects that finding out what is really the case and dealing with it is the best way to reproduce. Suppose that you are in the middle of the highway and a truck is bearing down on you. Thinking that you are reading this book might be some consolation as you wend your way into the infinite, but it is not the best of all possible ways to conceive and raise babies. On the other hand, spotting the truck and getting out of the way is a pretty good strategy. Spotting the real, objective truck is what Darwinism promotes. Those who are blind or stupid or overly philosophical get wiped out by natural selection. This is not to say that there are no cases of deception. Psychologists delight in deceiving first-year undergraduates with all kinds of puzzles about reasoning and perception. But note that these deceptions usually involve precisely those situations in which we generally do not find ourselves in real life. It is notorious that there are various kinds of mental tricks that fool almost everyone when they are presented abstractly, but that when put into everyday terms pose no problems. *Modus ponens*, affirming the consequent ("If P is true, then Q is true. Q is true. Hence P is true.") is a fallacy that most students find quite difficult to grasp when given in bare symbolic form. But as soon as you cash it out ("If you cheat then you will pass. You passed. Hence you cheated."), it is easy to spot as fallacious. And just as importantly, if there are cases of ongoing deception, there are good biological reasons. And we can find these things out. Hume, for instance, showed that there are no such things as causes, only constant conjunctions. But there are good reasons for thinking that there are causes. The burnt child fears the fire. Why? Because fire causes burning!

The point is that we have certain touchstones – grass is green; other people's genitalia (factoring in sex and orientation and so forth) are a turn-on; jumping off the Golden Gate Bridge is detrimental to your health; two plus two really does equal four – that we can rely on. We have no reason to think that under normal circumstances we are being deceived. And so, natural selection – extending all the way to Darwinian evolutionary theory – can pick up from there. But what about the worry, one that is taken up by Plantinga, that we humans are all a bit like the prisoners in the metaphor of the cave, introduced by Plato in his dialogue *The Republic*? Plato's prisoners are people in a cave, unable to get out, and looking always at a wall on which shadows of puppets are being projected. They think that the shadows are reality. But we know that they are not. In Plato's metaphor, someone escapes and finds reality outside the cave, even though when he returns the prisoners do not believe him and want to kill him. (The reference is to Plato's teacher, Socrates, who was put to death by the Athenian authorities for stirring up the philosophical imaginations of the young men of the city.) Could it not be that we are in the same position as the prisoners, and that "grass is green" is a shadow? Truly we know no better, and can never (at least, as long as Darwinism rules supreme) know any better. In the Darwinian case, there can be no possibility of escape and finding true reality. There can be no Socrates here.

All of which is undoubtedly true, but when you think about it, rather less troublesome than it appears to Plantinga and others who find this kind of radical scepticism convincing. We are not now talking about deception in any normal sense, but rather in some kind of metaphysical sense. We are not talking about thinking you have five dollars in your wallet when really you have only one. We are talking about thinking you have a real five dollars in your pocket when – even though you can pull out a five-dollar bill and pay for your coffee – in a God's-eye view you have nothing, or perhaps a ten-dollar bill. It is all a bit like that other philosophical favorite where we are all supposed to be, not living human beings, but brains in vats, wired up to get sensations of reality. This picture looks good, it seems to make sense, but truly it is all a bit of a sham. It is not what it seems to be. In our world, we are not brains in a vat, and that is all there is to it. Doctor Frankenstein had brains in a vat, and his story is rather scary only because this is so unusual. In any normal sense of the world, any normal understanding, we are not brains in a vat. Metaphysical scenarios are just that – metaphysical scenarios. There is no reason to think them reasonable or troublesome to us in our usual thinking about right and wrong, true and false, objective and subjective.

But is this not in the end to appeal to some kind of pragmatism? Truth is what works. Is this not to appeal to some kind of coherence theory of truth rather than a correspondence theory? Truth is what all hangs together rather than what is isomorphic to metaphysical reality. We are not prisoners in the cave because we can make sense of things by making all of our regular beliefs hang together. Metaphysical reality is nothing to us. We cannot use it. Likewise, we are not brains in a vat because everyday experience works and does not have wires and chemical baths and containers peering around the edges making the brain-in-the-vat hypothesis plausible. We can live without the hypothesis, so dismiss it. Probably there is indeed a kind of pragmatism and coherence here (Ruse 1998, 1999b). Plantinga objects to this, but really his objection is too much. Take up his own God hypothesis, without necessarily arguing whether or not God really exists. It is just not the case that someone who says we can never really tell if we are Plato's prisoners, brains in a vat, nevertheless can have a secure hot line to God, who will guarantee truth about the world. In the end, Plantinga is caught in the same pragmatism and coherence loop as the rest of us. There is no more reason to think that God's hot line is secure as there is to think that we are not prisoners. Plantinga's thinking here parallels that of René Descartes in the *Meditations*. The great French philosopher also thought he could get to God, without any possibility of error, and that from then on all properly derived knowledge claims were given the divine stamp of approval. Unfortunately, for all Descartes' subtle genius, no one else has ever thought his argument worked, and there is no reason to think that it is valid now for Plantinga. Like all viruses, radical scepticism strikes down believer and unbeliever without favor.

I stress again, I am not now saying that a Darwinian cannot be a Christian. I am saying that the charges that Darwinism collapses into refuting circularity are not well taken.

Darwinian Ethics

Let us now pick up on the morality side of philosophical inquiry. Just as both traditional evolutionary epistemology and evolutionary ethics fall for similar reasons, so both Darwinian epistemology and Darwinian ethics succeed for similar reasons. In particular, as in the case of epistemology, the secret to getting an adequate philosophy of morality is to take real Darwinism literally. Let us start at the level of substantive ethics, with questions about what one should do. If one takes seriously the division between fact and value – and I do – it will be realized at once that one cannot, from any amount of Darwinian biology, go straight to what one should do. The best one can do is go to what people think that they (and everyone else) should do. So let us start with that, and with the fact that if indeed Darwinian biology does stress the individual – at the ultimate, in the memorable metaphor of Richard Dawkins, "selfish genes" – one has an immediate paradox. Morality seems to be a question of doing things for others, and of thinking that one ought to do things for others. How can this kind of altruism come about from selfishness?

Of course, the first thing to recognize is that because one has selfish genes, this does not at once imply that one has selfish people. To speak of selfish genes is to speak metaphorically, and to imply that the genes are working in such a way as to promote their own replication. The second thing to recognize is that this promotion might involve giving things to others. But how might it involve giving things to others? One possible solution would be if selection worked for the benefit of the group rather than the individual (Sober and Wilson 1997). In this kind of situation, it would not be so much selfish genes as selfish species, and here one would expect and find that organisms would help conspecifics. The trouble is that (as we have seen) except in rather rare and technical situations, group selection seems not to occur. It is too open to cheating. In a group situation, if one has two organisms, one helping others and the other helping just itself, the self-interested one will be more likely to survive and reproduce, and so it will be the one more likely to spread its genes. Selfishness will therefore outstrip and swamp altruism. So it does seem that one must stay with some form of individual selection. But here the kinds of models we met in earlier chapters seem pertinent. Almost certainly, some form of kin selection is at work between close relatives. Why does a mother spend so much time with her offspring, other than because by so doing she is improving her own reproductive chances? And almost as certainly, a form of (what biologists call) reciprocal altruism is also involved. You scratch my back and I'll scratch yours (Trivers 1971).

Is this going to lead to any recognizable form of morality? Many would argue that morality involves a kind of social contract, and while the Darwinian denies (more accurately, has no reason to suppose) a social contract in the traditional sense – a group of people getting together to agree to certain rules – it is pretty clear that the kind of substantive ethics to which Darwinian biology leads is an ethics very much akin to what one would expect to find in a social contract. The best-known contemporary exposition of the social contract theory, that offered by John Rawls in his 1971 book *A Theory of Justice,* is very much the kind of morality that one would expect from a Darwinian perspective.

Rawls himself acknowledges this fact. He agrees that the social contract was hardly an actual event, but was rather something that was brought on by the struggle for existence working on our innate biology. Rawls argues that for a position that he calls "justice as fairness." He argues that in order to be just, one ought to be fair. For Rawls, being fair does not necessarily mean giving everybody absolutely equal shares of everything. Rather, he invites us to put ourselves behind what he calls a "veil of ignorance," not knowing what position we might find ourselves occupying in society: whether we will be male or female, rich or poor, black or white, healthy or sick, or any of these things. Then he asks what position self-interest dictates as the best kind of society to find oneself in, and his answer is that it is a society where in some sense everybody does as well as one might possibly expect, given our various talents.

It may well be that we will be born male and rich and powerful and healthy and so forth. If we knew we were going to find ourselves in that position, then we would want maximally to reward people in that position. But, of course, we may be female and poor and helpless, in which case we would lose out. So there is a kind of initial presumption of equality. Because we do not know what position we will have in society, it is best to assume the worst and then to make sure that the worst does not do any worse than anyone else. Everyone should get the same. Yet this presumption that equal distribution is in one's interests is overthrown as soon as one recognizes that something like the availability of good medical care is going to be of benefit to everybody. If the only way that you can get the most talented people to become doctors is by paying them more than twice what you pay university professors, then "justice as fairness" dictates the propriety of this kind of inequality. Whatever your position in society, you would rather have – it is in your own self-interest to have – good medical care than absolutely equal salaries.

So what Rawls ends up with is a society with inequalities, but in some sense a society where the inequalities benefit each and every individual in the group. This is very much the kind of society that one expects evolutionary biology to have produced – that is to say, a group of people who think that one ought to be just, meaning that one ought to be fair. A group that recognizes that there will be inequalities, but that also recognizes that these inequalities will in some sense be of benefit to all. This is not to say that every actual society has turned out like this, but that is not really the point. One recognizes there are going to be all sorts of ways in which biology fails to match what the genes might dictate as best. There may be inequalities brought about by particular circumstances or fortune or whatever. But the point is that this is how we think that a society ought to be, even if it is not necessarily always that way. Although note how unstable are societies, like the Third Reich, that blatantly and egregiously violate rules of fairness. Even societies that we may (properly) judge as unfair in some overall sense – societies based on a slave system, for instance – usually flourish only if and inasmuch as everyone gets some kind of break. If you mistreat your slaves, you are going to get rebellion and worse.

And this point about societies – the distinction between what one might think morally ought to be, and what actually comes about – reflects down to the same point about individuals. Things actually come about the way they do because of individuals, but nobody is ever saying that people always act morally or follow the dictates of their conscience. Morally, it is a question of how you ought to behave, not of how you do behave. If, for instance, I find myself rich and powerful, then from a purely selfish point of view I might simply ignore the call of morality, or suppress it, or educate myself and my children to ignore our biology. I or any other Darwinian am not saying that this sort of thing might never happen. At times of great social stress, because of a natural disaster like a plague, one might well find the rich and powerful looking after themselves alone. As Christians remind us when they talk of original sin, the existence of a moral code does not preclude selfishness on the part of individuals, even individuals who accept the binding validity of the code.

There are many moral systems other than those focusing on a social contract. However, a Darwinian approach does not necessarily refute or repudiate what has just been said in favor of a social contract theory. Most moral systems agree on the basics and only come apart on the kinds of esoteric, artificial examples that philosophers delight in. Everybody, whether a Kantian or a utilitarian or a Christian or whatever, thinks that one ought to be kind to small children, and that gratuitous cruelty toward the aged is wrong, and that there is something to be said for honesty and decency in business and relationships, and so on and so forth. Most morality in fact is a fairly commonsense morality – be decent to each other, care about the unfortunate and weak, do not let cheats get away with it – rather than a well-articulated system as produced by philosophers (Mackie 1977). And this point is precisely what the Darwinian would insist on. Real human beings have a commonsense morality – just as they have a commonsense rule of reasoning – that guides them in their everyday life.

Game Theory

This is a naturalistic position on ethics (as it was on epistemology). If the science fails, then so does the philosophy. Have we any reason to think that – even if we agree that a Rawlsian type of situation could and would be maintained by selection – it has ever come into being? This is an option for intelligent agents, rather than beings that are under the control of the genes and hence might not be planning at all for themselves. There are various ways in which one might start to approach the empirical questions, most obviously by turning to evidence of moral or proto-moral behavior in other species. Much interest has been shown in our close relatives, the chimpanzees, especially the pigmy version (bonobos). Students of their behavior argue strongly that we do find actions strongly suggestive of cooperation that simulates the moral (de Waal 1982, 1996). Dogs also, being highly social, might fit in here (Figure 10.1).

Another approach to empirical validation focuses on game theory, that system of applied mathematics that assesses strategies in situations of competition. Models are now showing that some kind of justice-like reciprocation can evolve, even when no planning is involved at all (Skyrms 1998).

PICKLES by Brian Crane

FIGURE 10.1. "Pickles" cartoon by Brian Crane (January 8, 2005). © 2005 The Washington Post Writers Group. Reprinted with permission.

To see this, let us introduce two important concepts (one of which we have met before). The first is the notion of a Nash equilibrium. This is a kind of situation where, given two players in a game, if they are fighting over a fixed sum, and if they together demand more than the sum, neither will get anything. Given that both players know what the other will do, what is the most rational move for this first player? Suppose, for instance, that there are 100 units to be divided and that player 1 knows that player 2 will demand 70 units. Then the most rational demand for player 1 is 30 units. An equilibrium holds if the distribution is 30:70 – player 1 cannot do better than this, and could do worse. The second notion is that of an evolutionarily stable strategy; this is a situation, it will be remembered, where no one mutant or variation can gain over a certain ratio against others in the population. Selection for rareness will lead to such an equilibrium, because if the variation gets more common, it will be under heavier selection pressure, and conversely.

Now fairness would seem to demand that the two players agree to divide 50:50, but why should this evolve, given that it could be rational to go 30:70 given the greediness (but not irrationality) of player 2? The philosopher Brian Skyrms writes as follows:

> Suppose we put this game in the evolution context that we have developed. What pure strategies are evolutionarily stable? There is exactly one: Demand half! First, it is evolutionarily stable. In a population in which all demand half, all get half. A mutant who demanded more of the natives would get nothing; a mutant who demanded less would get less. Next, no other pure strategy is evolutionarily stable. Assume a population of players who demand x, where $x < \frac{1}{2}$. Mutants who demand $\frac{1}{2}$ of the natives will get $\frac{1}{2}$ and can invade. Next consider a population of players who demand x, where $x > \frac{1}{2}$. They get nothing. Mutants who demand y, where $0 < y < (1 - x)$ of the natives will get y and can invade. So can mutants who demand a $\frac{1}{2}$, for although they get nothing in encounters with natives, they get $\frac{1}{2}$ in encounters with each other. Likewise, they can invade a population of natives who all demand 1. Here the symmetry requirement imposed by the evolutionary setting by itself selects a unique equilibrium from the infinite number of strict Nash equilibria of the two-person game. The 'Darwinian Veil of Ignorance' gives an egalitarian solution. (Skyrms 2002, 276)

Suppose we grant all of this? You may still complain, legitimately, that we do not have morality. We have beings behaving as if they were moral. Morality, however, involves a sense of moral obligation – I not only feed the sick and poor, I feel that I ought to feed the sick and the poor. At this point, obviously, the Darwinian ethicist supposes – that is, makes an empirical assumption – that this sense of obligation is something put in place by selection in order to make us work together, to make us altruists who respect fairness. Normally, we are self-centered. That is the way that selection has made us. Normally, selfish genes produce selfish people. So we look to our own needs when it comes to food and sex and so forth. But we are social animals also, and there are advantages to being social. So we have this moral sentiment to make us reach beyond ourselves. Morality in this sense is an adaptation, just like any other.

Again one must stress that this is a naturalistic position, but again one can note that it is not without support. Work is now proceeding at an empirical level showing how moral-type sentiments emerge in games of strategy (Frank 1988, 1999). But at the general level, the most obvious empirical support for the suggestion that ethics (substantive ethics) is an adaptation is that it fits in with the general Darwinian picture. We do have biological inclinations to selfishness – we want food and mates for ourselves – and so, if cooperation is of value (and we saw in an earlier chapter that this is the way of human evolution), we need adaptations that will let us break through the selfishness. A moral sense is just what is needed. And note that it is a real moral sense. You might think that it could all be a matter of self-calculation, but apart from the fact that this is simply not true – people do do things because they are right and not just because they are selfish – self-interested calculations have a cost. In life, time is money. If you are deciding whether to help someone – save someone from a tiger – by the time you have made the calculation, it may be too late – the tiger has eaten both of you. Substantive ethics is a kind of quick-and-dirty solution to the question of cooperation. It gets you to act quickly, even though (as with quick-and-dirty solutions) it might not always be the best answer.

Metaethics?

Note that the Darwinian ethicist does not merely acknowledge the difference between claims of fact and claims of obligation, but also stresses that this difference is real. It is precisely because we have this special strong kind of emotion – one that can make us act when other emotions are pointing the other way – that cooperation works as well as it does. So what price justification? It is one of life's supreme ironies that people at opposite ends of the spectrum so often come together in criticizing or belittling the middle – Wittgenstein and Popper on Darwinism, for example. Likewise with the Christian philosopher Alvin Plantinga and the secular philosopher Philip Kitcher. The latter has nothing but contempt for the former, going so far as to say that Plantinga's feelings about origins puts him in the company of those who have "put their brains in cold storage" (Kitcher 2001, 261). Yet when it comes to Darwinian approaches to human nature, including human thought systems, Kitcher (2003) shows himself no less critically scathing than Plantinga. Of course, Kitcher hardly objects to the sorts of normative claims that are being made in the name of Darwinism. But even if he would grant that Darwinism itself has any real place in explaining the existence of such sentiments, he thinks that any attempt to go further – to relate substantive ethics to foundational questions (through Darwinism) – is a straight "blunder." Indeed, "we should dismiss this endeavor as deeply confused" (p. 327).

Now, if it were the case that the Darwinian ethicist was simply trying to get foundations in the same way as the traditional evolutionary ethicist, there would be reason for these sneers. You cannot go from the course or fact of evolution to moral prescriptions. We have seen this. But the Darwinian ethicist will not attempt this. It was just emphasized that the Darwinian ethicist wants to reaffirm the validity of the is/ought distinction. To the Darwinian, the naturalistic fallacy is genuinely fallacious. So what can be the move of the Darwinian? To take a radically different approach and to deny that there are any foundations at all! The Darwinian's answer to the question of justification is that ethics – substantive ethics, that is – has no justification (Mackie 1979; Murphy 1982). This is not to deny that we humans think it has a justification – that we think it objective – but this is part of its adaptive nature. If we thought that ethics were simply emotion, it would soon break down as people realized that it was not binding and started to cheat. So natural selection leads us to think that we ought to do things, not just from emotion or desire, but because they are "really and truly right." But in fact, substantive ethics has no referent. It is just a bunch of fancy emotions. In this sense, ethics is an illusion – an adaptation put in place by our genes to make us cooperators. Substantive ethics as such is not illusory, but the belief that it is objective is illusory.

Now, how can one argue this? Obviously by pointing out that ethics (substantive ethics) is an adaptation and by concluding that, from this fact, it can have no objective referent. It is here that Kitcher – following in the tradition of others, like the philosopher Robert Nozick (1981) – strikes. "That this reasoning is fallacious is evident once we consider other systems of human belief. Plainly we have capacities for making judgments in mathematics, physics, biology, and other areas of inquiry. These capacities too have historical explanations, including, ultimately, evolutionary components" (Kitcher 2003, 323). But, implies Kitcher, these other areas of inquiry are clearly objective – they are about reality in some sense – so how can we deny this to morality, simply on the grounds of the evolved nature of our moral sentiments?

Given that this objection is so obvious, one queries whether the Darwinian ethicist can be making quite so naïve an argument? Hardly. We go back to the question of progress, or rather to the nonprogressive nature of the evolutionary record. If there is no progress, then there was no inevitability that substantive ethics would have evolved as it did – unless, of course, there was something objective out there drawing it, or unless the only way in which cooperation could have been achieved was through substantive ethics as we know it. (The former solution would be akin to that of someone like Plato, who thought that there are transcendent values, and the latter to someone like Kant, who thought that ethics is a condition of rational people interacting socially.)

Let us agree that perhaps the game theorists can show us that cooperation is going to demand formal interactions along the lines that we do have today, thanks to our substantive ethics. In other words, that these rules of engagement are inevitable. This does not give us substantive ethics, as Kant (1959) realized (for all that he was aiming to get substantive ethics from a formal argument). You need, as it were, to fill in the gaps between the symbols, to show how one is motivated to follow the formal rules. And our way – love your neighbor as oneself (or justice as fairness, or some such thing) – is by no means the obviously unique way. Consider what one might call the "John Foster Dulles system of morality." Dulles, the secretary of state under Eisenhower during the Cold War, hated the Russians – he thought them wrong, and he believed that he should hate them. He realized also that the Russians hated him, in the same moral way. Hence, he and the Russians cooperated. Here we have a system where the supreme moral imperative is to hate others; you have an obligation to hate others, but remember that they have the same feelings and obligations toward you. So you get on together.

Two conflicting moral systems – our own and the Dulles version. If evolution were progressive, we could pick out ours as the right one, but evolution is not so obviously progressive, so in the absolute scheme of things there is nothing to choose between them. Which means that, at the very least, if there is more to ethics than emotion, we could be living a total lie – not just about the belief in foundations but about substantive ethics itself. At the very least, one can say that this is not what people usually mean by objective foundations to ethics – for a Platonist, there is something contradictory in thinking that an objective foundation exists but has no effect on humans. Or, if you insist that an objective foundation could still exist, even though we might be unaware of it – a kind of parallel position to that envisioned by Plantinga, where an objective world could exist though we are ignorant of it – we can still say that as far as we humans are concerned (paralleling the reply to Plantinga), what matters is our biological past and how we make do with it today, rather than the dictates of something beyond our possible ken.

Enough! Wittgenstein was wrong. Darwinism is highly relevant to philosophy – as anyone but a professional philosopher would expect. Popper was wrong. Darwinism is a scientific theory. It is not a philosophical theory. But it has a lot to tell us about philosophy.

CHAPTER ELEVEN

Literature

There is a hideous fatalism about it, a ghastly and damnable reduction of beauty and intelligence, of strength and purpose, of honor and aspiration, to such casually picturesque changes as an avalanche may make in a mountain landscape, or a railway accident in a human figure. To call this Natural Selection is a blasphemy, possible to many for whom Nature is nothing but a casual aggregation of inert and dead matter, but eternally impossible to the spirits and souls of the righteous.

G. B. Shaw (1988)

George Bernard Shaw, the Irish playwright, never minced words. His judgment of Darwin's theory of evolution was unequivocally negative. "What damns Darwinian Natural Selection as a creed is that it takes hope out of evolution, and substitutes a paralysing fatalism which is utterly discouraging. As Butler put it, it banishes Mind from the universe." Let us start with Butler – Samuel Butler, the late Victorian novelist – then move to Shaw and others.

Erewhon

Samuel Butler's grandfather was the headmaster of Shrewsbury School when Darwin attended the school as a pupil. His real connection to our story is less tenuous, and comes from his lifelong interest in evolutionary ideas. Were he not a writer of fiction, he would still warrant our attention. He became one of Darwin's most bitter critics, and representative of many who wanted to be evolutionists but who hated what they saw as the blind materialism of the Darwinian mechanism of natural selection. His was a classic Victorian life: he intended to become an Anglican clergyman, lost his faith in Christianity, took up other work – in his case, somewhat enterprisingly, sheep farming for a while in New Zealand – and spent the whole of his days trying to find some meaningful substitute for his nonbelief (Holt 1964). In the late 1870s and into the 1880s, having persuaded himself that Darwin was not merely wrong but a charlatan, Butler (1879) penned a series of books that laid out a kind of purposeful evolution – in some sense guided upward and yet not God-controlled – that had the Lamarckian use and disuse and inheritance of acquired characteristics at its causal center. At the same time he was working on his most important novel, *The Way of All Flesh* – unpublished until after his death in 1902 – which uses ideas of recapitulation to explain the ways in which family characteristics keep reappearing over the generations.

What commands Butler to our attention is the book that made him first famous, his brilliant satire *Erewhon* ("nowhere" spelled backwards), which appeared in 1872. This is a clever attack on those reformers who were arguing that crime and misbehavior are not moral failings but signs of deprivation and sickness. The narrator visits a far-off land, where those who are ill are incarcerated but those who are evil are coddled and taken to hospital to recover. This strange world, *Erewhon*, has no place for machines, and in perusing a copy of an old manuscript – "The Book of the Machines" – the narrator finds out why. (This part of the novel was written separately, before the rest of the novel, and first published in a newspaper in Christchurch, New Zealand, in 1863, after Butler had read the *Origin*.) Machines had been banned because the authorities feared that they might evolve and eventually take over the world, reducing humans to the status of second-class citizens. But could this possibly be so? We have consciousness, and machines do not. However, points out the author of the manuscript, if evolution be true (and the assumption is that it is true), then humans came from beings that did not think and reflect.

"There is no security" – to quote his own words [the author of the manuscript] – "against the ultimate development of mechanical consciousness, in the fact of machines possessing little consciousness now. A mollusc has not much consciousness. Reflect upon the extraordinary advance which machines have made during the last few hundred years, and note how slowly the animal and vegetable kingdoms are advancing in comparison. The more highly organised machines are creatures not so much of yesterday as of the last five minutes, so to speak, in comparison with past time. Assume for the sake of argument that conscious beings have existed for some twenty million years: see what strides machines have made in the last thousand! May not the world last twenty million years longer? If so, what will they not in the end become? Is it not safer to nip the mischief in the bud and to forbid them further progress?" (Butler 1872, 191–2)

Machines like organisms? Future machines like possible humans with the power of thought? Butler gives a brilliant defense of this possibility, showing that all of the possible disanalogies between organisms and machines are no true bar to the future mechanical thinkers. In respects, machines are already like organisms. As importantly, in some respects organisms, including humans, are already like machines. Think of the eye. What is this but a machine for seeing? More than this, we can say already that we exist for machines as much as machines exist for us. When a man picks up a spade to dig, the spade exists for our benefit. But in a converse way, we exist for the spade – it could not function were we not around, and we play a machine-like role for the spade. Our stomachs and powers of digestion exist for the spade no less than the energy-producing powers of a complex machine exist for us.

Thus follows the argument that machines could evolve to consciousness and take over the world. Is there no counterargument? The end of the manuscript brings in one who dissented from this gloomy prospect, defending machines on the grounds that they are really but an extension of human nature and our powers.

"Its author [the defender of machines] said that machines were to be regarded as a part of man's own physical nature, being really nothing but extra-corporeal limbs. Man, he said, was a machinate mammal. The lower animals keep all their limbs at home in their own bodies, but many of man's are loose, and lie about detached, now here and now there, in various parts of the world – some being kept always handy for contingent use, and others being occasionally hundreds of miles away. A machine is merely a supplementary limb; this is the be all and end all of machinery. We do not use our own limbs other than as machines; and a leg is only a much better wooden leg than any one can manufacture. In fact, machines are to be regarded as the mode of development by which human organism is now especially advancing, every past invention being an addition to the resources of the human body. Even community of limbs is thus rendered possible to those who have so much community of soul as to own money enough to pay a railway fare; for a train is only a seven-leagued foot that five hundred may own at once." (221)

Critique of Darwinism?

There are several ways in which one could read all of this. Most obviously, it can be read as a staggeringly prescient account of the ways in which machines can and do develop – although Butler knew of elementary calculating machines, all of this was written a hundred years before computers started to take over our lives. (Interesting metaphor!) A second reading takes the story as an attack on Darwinism. Butler acknowledged in a Preface to a later edition (1901) that many had taken it this way. It is obviously absurd to think that machines could ever evolve to an intelligent state. Hence, it is obviously absurd to think that humans, starting with life in a primitive, unthinking state, could have evolved into their present thinking form. The story of the machines is to be regarded as an elaborate reductio ad absurdum of the whole Darwinian story, especially as it applies to humans.

Butler denied that this was his intent, and I think we may take him at his word. "I regret that reviewers have in some cases been inclined to treat the chapters on Machines as an attempt to reduce Mr. Darwin's theory to an absurdity. Nothing could be further from my intention, and few things would be more distasteful to me than any attempt to laugh at Mr. Darwin; . . ." (Butler 1901). Especially today, the inclination is to read the story as sketching a possibility rather than an impossibility, without feeling the need to work at showing pertinent disanalogies between machines and people. If it is a possibility, then that is no refutation of Darwinism.

A third reading sees in Butler's story a basis for an improved evolutionary theory, a rival to traditional Darwinian selection-based change. Butler himself saw things this way, especially in the rival analysis of machines that he offers at the end of the story. He thought that this was the basis for his own theory of change through habit and Lamarckian inheritance (Butler 1879). Somehow we change things in our lifetime, and then the changes get passed on. In the machine case, we improve machines – which are in a very real sense an extension of us – and then these get passed on directly to the next generation. So likewise for us, whose bodies can be regarded as machines. We improve ourselves through our own actions and desires, and then these better-quality features get transmitted directly to the next generation.

To which we can say that although Samuel Butler was a biting critic of Victorian society, he was no deep challenger to Darwinism. For a start,

Lamarckism is false – in the biological world, there is no inheritance of acquired characteristics. For a second, following this – in a manner akin to the traditional evolutionary epistemologists of the last chapter – Butler is blurring or ignoring the prime distinction between human innovation and biological variation. The former is produced to order, the latter is not. Even if we agree that there is progress in science and technology, there is as yet no basis for saying that there is progress in the world of organisms. Culture is directed. Organic change, unless powered by human invention, is not. Butler does not refute Darwinism, although his thinking does explain why (in his autobiography) Karl Popper, having said that he got no insight into the nature of evolutionary change from evolutionists after Darwin, then pulled himself up and named one who did meet with his approval: Samuel Butler!

George Bernard Shaw

Shaw, the great Irish playwright, liked Butler also. He loathed and detested Darwinism. Darwinism is blind, vicious, and materialistic. It justifies war and unbridled competition. It does not lead to progress. It cannot lead to progress. Natural selection is just a blind process, based on struggle and cruelty, without any purpose or forward motion. Darwinism does not even have the virtue of originality, for it is simply the continuation of a gross political philosophy that had taken root in the eighteenth century.

> Long before Darwin published a line, the Ricardo-Malthusian economists were preaching the fatalistic Wages Fund doctrine, and assuring the workers that Trade Unionism is a vain defiance of the inexorable laws of political economy, just as the Neo-Darwinians were presently assuring us that Temperance Legislation is a vain defiance of Natural Selection, and that the true way to deal with drunkenness is to flood the country with cheap gin and let the fittest survive. (Shaw 1988)

Indeed, Darwinism is so stupid an idea, one wonders – except for blatant self-interest – why anyone would take it seriously. The clue lies in the fact that it is such a trivially simple idea that it seduces people into thinking that they have grasped great truths and insights. It is nothing of the sort.

In 1906, referring to Darwin, Huxley, Spencer, and their friend and supporter John Tyndall, Shaw wrote:

> I really do not wish to be abusive; but when I think of these poor little dullards, with their precarious hold of just that corner of evolution that a blackbeetle can

understand – with their retinue of twopenny-halfpenny Torquemadas wallowing in the infamies of the vivisector's laboratory, and solemnly offering us as epoch-making discoveries their demonstrations that dogs get weaker and die if you give them no food; that intense pain makes mice sweat; and that if you cut off a dog's leg the three-legged dog will have a four-legged puppy, I ask myself what spell has fallen on intelligent and humane men that they allow themselves to be imposed on by this rabble of dolts, blackguards, impostors, quacks, liars, and, worst of all, credulous conscientious fools.

If this was Shaw trying not to be abusive, one wonders what it would be like if he did try. Some good idea can be gleaned from the play *Back to Methuselah*, which he wrote after the First World War, a catastrophe Shaw laid at the feet of Darwinism. The lengthy Preface to the play was a diatribe against the vile militaristic and capitalistic philosophy that Shaw attributed almost exclusively to the English naturalist.

At the present moment one half of Europe, having knocked the other half down, is trying to kick it to death, and may succeed: a procedure which is, logically, sound Neo-Darwinism. And the goodnatured majority are looking on in helpless horror, or allowing themselves to be persuaded by the newspapers of their exploiters that the kicking is not only a sound commercial investment, but an act of divine justice of which they are the ardent instruments. (Shaw 1988)

The play itself was an attempt to provide a kind of romantic text or fable that would adorn and supplement and make more convincing and attractive the kind of Creative Evolutionism that he (following Butler) preferred. We need a kind of new religion, and he was about to provide it.

Shaw's own Creative Evolution – which certainly bears strong resemblances to Bergsonian vitalism as much as to anything home-grown – depends on a kind of willing for change combined with a Lamarckian use and disuse backed by the inheritance of acquired characteristics. We humans are striving to improve things, and this will ultimately lead to a progression upward. The play itself is intended to illustrate this point. Divided into five parts (with a biblical echo, it is subtitled "A Metabiological Pentateuch"), we are taken from the Garden of Eden to some point in the very far future (31,920 A.D.). In the first part, set back in the Garden of Eden, discovering the possibility of accidental death, Adam decides voluntarily to limit his life span. "I am tired of myself. And yet I must endure myself, not for a day or for many days, but for ever. That is a dreadful thought." Yet how can this be changed, and how can life continue if Adam no longer exists? How can one guard against both boredom with oneself and the possibility of accidental death that brings all life to an end? Through the birth of a new generation and the death of the old. Can this be done? Yes, says the serpent. All changes are possible through the will and imagination. "You imagine what you desire; you will what you imagine; and at last you create what you will." So Adam makes his resolution: "I will live a thousand years; and then I will endure no more: I will die and take my rest."

The rest of the play follows human history as it takes its course upward toward immortality – a kind of disembodied state of being superior to that which so burdened Adam, and guarded against accidental death. This path is taken in steps. By the beginning of the twentieth century, life has degenerated to a span of but a few decades. First, we must go back to the span of the figures of the early books of the Bible. (Hence the title of the play, for Methuselah was reputedly the oldest of men, living for nearly a thousand years.) An act of will is needed to extend our expectations to three hundred years, and then more, and more. Finally, we can be really long-lived and ready for the next step to immortality. The fifth and final part of *Back to Methuselah* shows the climax of the evolutionary process. By now humans are born, from eggs, at about the age of seventeen in our terms, all of childhood and adolescent development condensed down to rapid early stages. For four years, we lead lives of carefree happiness – with lots of sex. Then, at about that age, sexual and social desire starts to end – breasts wither and hair falls out.

THE YOUTH. You old fish! I believe you don't know the difference between a man and a woman.

THE ANCIENT. It has long ceased to interest me in the way it interests you. And when anything no longer interests us we no longer know it.

We become potentially immortal, grown beings with interests all and only at the conceptual level.

THE MAIDEN [aged about four years and starting to lose interest in the habits of childhood]. What does it matter what I did when I was a baby? Nothing existed for me then except what I tasted and touched and saw; and I wanted all that for myself, just as I wanted the moon to play with. Now the world is opening out for me. More than the world: the universe. Even little things are turning out to be great things, and becoming intensely interesting. Have you ever thought about the properties of numbers?

THE YOUTH [*sitting up, markedly disenchanted*] Numbers!!! I cannot imagine anything drier or more repulsive.

THE MAIDEN. They are fascinating, just fascinating. I want to get away from our eternal dancing and music, and just sit down by myself and think about numbers.

Still there is the possibility of life ending (as for Adam) through an accident, but eventually all will dissolve into pure thought and immortality.

THE SHE-ANCIENT. But still I am the slave of this slave, my body. How am I to be delivered from it?

THE HE-ANCIENT. That, children, is the trouble of the ancients. For whilst we are tied to this tyrannous body we are subject to its death, and our destiny is not achieved.

THE NEWLY BORN. What is your destiny?

THE HE-ANCIENT. To be immortal.

THE SHE-ANCIENT. The day will come when there will be no people, only thought.

THE HE-ANCIENT. And that will be life eternal.

One can never be certain that Shaw is not poking fun at people, including himself. His scientific theory, if one can dignify it with such a term, is nonsense. It is in some sense recapitulatory, with the whole of life (as we more primitive folk enjoy it) collapsed or condensed into the first few years, and then with old age extended out indefinitely – old age, without Alzheimer's and all of the other infirmities that plague us now. To this is added some kind of vitalistic impulse to life, which leads upward. Much more interesting is Shaw's vision of how a progressive form of life might climax, however caused. We know how, even today, many find it hard to accept Darwinism simply because it does not guarantee the kind of progress that they want – they feel instinctively that it cannot be true because it is so random and undirected. It is worth asking precisely what kind of progress it is that people do want. Do people want the climax that Shaw sketches? I suspect that many would say that they do – but that as soon as his sketch is filled out, they find the reality so repellent that they want no part of it!

Does one really want a life, even an eternal life devoted to the joys of mathematics, if there is no love and no sex and no food (certainly no interest in food) and no anything else that we humans find so worthwhile – including the joys and troubles of having children? Do we really want to be superbrains that have transcended the physical and live only in the mental? Do we really want to live forever in this fashion? Far better to be one of Shaw's infants than to be one of his old-timers. Of course, the response is that that is precisely what one would expect of infants – if they knew and could appreciate the joys of old age, they would change their minds. "Better Socrates dissatisfied than a fool satisfied." Perhaps so, but we humans today can judge only as infants, and we do know the pleasures and pains of our lives. Do we really want to swap them for the Shavian better state?

Paradoxically, Shaw shows that it might not be such a bad thing that good old-fashioned Darwinian evolution does not point to the inevitability of Shaw's future human race. Those who reject Darwinism because it does not support their vision of the future might first stop and ask themselves what kind of future they really want.

Social Darwinism

Not every writer using evolutionary themes has been a progressionist. H. G. Wells, a student of T. H. Huxley, writing toward the end of the nineteenth century, used fiction to express a gloomy picture of the future. In his science fiction novel *The Time Machine*, Wells supposes that the prospects for the human race are anything but rosy. A traveler to the future finds that there are two species, one (Eloi) living above ground, having little intellectual curiosity about anything, and the other (Morlocks) more inventive and intelligent, living in subterranean caves, repulsive brutes who eat the above-ground dwellers. "I felt a peculiar shrinking from those pallid bodies. They were just the half-bleached color of the worms and things one sees preserved in spirit in a zoological museum. And they were filthy cold to the touch" (Wells 1895, 111).

And yet, realizes the traveler, we have to accept that Morlocks, no less than Eloi, are descended from us.

> I do not know how long I sat peering down the well. It was not for some time that I could succeed in persuading myself that the thing I had seen was human.

But, gradually, the truth dawned on me: that Man had not remained one species, but had differentiated into two distinct animals: that my graceful children of the Upper World were not the sole descendants of our generation, but that this bleached, obscene, nocturnal Thing, which had flashed before me, was also heir to all the ages. (p. 101)

Others have had a more inspiring picture of the end of progress than did Wells. In a very popular novel of the fourth decade of the last century, *Last and First Men: A Story of the Near and Far Future*, the science fiction writer Olaf Stapledon supposes that we are going to evolve into beings with a kind of ability to form a group mind. In some way, our successors – the "Last Men" – have found a method of communicating with each other telepathically. They can come together to form just one brain rather than many.

Occasionally there is a special kind of group intercourse in which, during the actual occurrence of group mentality, all the members of one group will have intercourse with those of another. Casual intercourse outside the group is not common, but not discouraged. When it occurs it comes as a symbolic act crowning a spiritual intimacy.

Unlike the physical sex-relationship, the mental unity of the group involves all the members of the group every time it occurs, and so long as it persists. During times of group experience the individual continues to perform his ordinary routine of work and recreation, save when some particular activity is demanded of him by the group mind itself. But all that he does as a private individual is carried out in a profound absent-mindedness. In familiar situations he reacts correctly, even to the extent of executing familiar types of intellectual work or entertaining acquaintances with intelligent conversation. Yet all the while he is in fact 'far away', rapt in the process of the group-mind. Nothing short of an urgent and unfamiliar crisis can recall him; and in recalling him it usually puts an end to the group's experience. (Stapledon 1930, 319)

There is no conflict between individuals except through mistakes. Together, these men achieve something that would be impossible were they to think and work on their own. "The only kind of conflict which ever occurs between individuals is, not to the irreconcilable conflict of wills, but the conflict due to misunderstanding, to imperfect knowledge of the matter under dispute; and this can always be abolished by patient telepathic explication" (pp. 320–1).

Stapledon is trying to picture a society where humans (or their successors) do not fight, but live in harmony. Obviously this is set against the background belief that Darwinism normally leads to conflict and pain and loss – although, given that Darwin himself drew on earlier sources, Malthus and Lyell, creative thinkers had, not surprisingly, picked up on this theme even before the *Origin* was published. Famously, in his poem "In Memoriam," Alfred Tennyson asked:

> Are God and Nature then at strife,
> That Nature lends such evil dreams?
> So careful of the type she seems,
> So careless of the single life;
>
> . . . .
>
> So careful of the type? but no.
> From scarped cliff and quarried stone
> She cries, 'A thousand types are gone:
> I care for nothing, all shall go.'

Tennyson found an answer in a Butler-like creative process leading on to higher types and thus justifying or mitigating the suffering experienced on the way to the climax (Ross 1973; Ruse 1999a). But after Darwin, expectedly, there were those who mirrored in fiction the theorizing of the (best-known) Social Darwinians, using the struggle as the power behind a kind of beneficent selective force that leads to progress. This was a common theme in naturalistic American novels at the beginning of the twentieth century – in *The Octopus*, for instance, a novel by Frank Norris about the struggle in California between the wheat growers and the all-encroaching railroad company. The hero (Presley) meets the baron of the railroad (Shelgrim), a man who is ruthlessly forcing the tracks across the state, irrespective of the lives and happiness of others (represented here by the farmer Derrick).

> "And," continued the President of the P. and S. W. with grave intensity, looking at Presley keenly, "I suppose you believe I am a grand old rascal."
>
> "I believe," answered Presley, "I am persuaded – – " He hesitated, searching for his words.
>
> "Believe this, young man," exclaimed Shelgrim, laying a thick powerful forefinger on the table to emphasize his words, "try to believe this – to begin with – *that railroads build themselves*. Where there is a demand sooner or later there will be a supply. Mr. Derrick, does he grow his wheat? The Wheat grows itself. What does he count for? Does he supply the force? What do I count for? Do I build the Railroad? You are dealing with forces, young man, when you speak of

Wheat and the Railroads, not with men. There is the Wheat, the supply. It must be carried to feed the People. There is the demand. The Wheat is one force, the Railroad, another, and there is the law that governs them – supply and demand. Men have only little to do in the whole business. Complications may arise, *conditions that bear hard on the individual – crush him maybe – but the wheat will be carried to feed the people* as inevitably as it will grow. If you want to fasten the blame of the affair at Los Muertos on any one person, you will make a mistake. Blame conditions, not men."

"But – but," faltered Presley, "you are the head, you control the road."

"You are a very young man. Control the road! Can I stop it? I can go into bankruptcy if you like. But otherwise if I run my road, as a business proposition, I can do nothing. I can not control it. It is a force born out of certain conditions, and I – no man – can stop it or control it. Can your Mr. Derrick stop the Wheat growing? He can burn his crop, or he can give it away, or sell it for a cent a bushel – just as I could go into bankruptcy – but otherwise his Wheat must grow. Can any one stop the Wheat? Well, then no more can I stop the Road." (Norris 1901, 576)

Determinism

Obviously Social Darwinian; less obviously Darwinian. As with the philosophy, the connection here with Darwin's theory of evolution is (shall we say) somewhat loose. A far greater influence is Herbert Spencer. Consider the closing lines of the novel. Lives have been broken – death, poverty, crime, insanity, prostitution. But the railroad is built, and the wheat grows, all for the ultimate benefit of humankind.

Falseness dies; injustice and oppression in the end of everything fade and vanish away. Greed, cruelty, selfishness, and inhumanity are short-lived; the individual suffers but the race goes on. Annixter [a character in the novel] dies, but in a distant corner of the world a thousand lives are saved. The larger view always and through all shams, all wickednesses, discovers the Truth that will, in the end, prevail, and all things, surely, inevitably, resistlessly work together for good. (651–2)

Note the underlying assumption of determinism or fatalism – no matter what any individual may do, nothing can stop the march of progress. In the novel, actually, Shelgrim is presented as a man of some sensitivity, not to say moral understanding. (Instead of firing a drunken worker, Shelgrim raises his salary.) He is also being used to make explicit the notion that human intentions are essentially irrelevant, that what will be will be regardless of our desires and hopes. For Norris, this is the real import of evolution – human society is ruled by the uncaring laws of nature no less than a rock or a stone. "Men have only little to do in the whole business." At this point we link to Butler and Shaw, for they agreed that even more significant than the violence and its supposed justification was the imputation of inevitability. They too saw Darwinism as implying determinism, fatalism, over which we have no control. Remember: "There is a hideous fatalism about it, . . ." and "it takes hope out of evolution, and substitutes a paralysing fatalism." People are like machines, and there is no place for freedom or choice.

It was this omission that Creative Evolution was intended to remedy. Through a kind of Lamarckian effort, our labors would have lasting results; through acts of will, evolution was to be directed according to human desires. Where Butler and Shaw differ from Norris, or at least from his mouthpiece in *The Octopus*, is that whereas they are repelled by the prospect of nature's control, Shelgrim seems excited, or at least resigned. For this reason, whereas Butler and Shaw want to replace Darwinism, Norris (however genuinely Darwinian we judge his thinking) has no such desire. But different reactions notwithstanding, the premise of determinism or fatalism is shared.

This is a question to which we must pay attention. Does Darwinism truly deny human freedom; does it make the actions of the individual unimportant? After the *Origin* – more particularly, after the *Descent* – are humans little more than fancy machines (as Butler suggests), or is there still room for human choice and action? Are we autonomous – morally autonomous – or are we mere robots? Butler, Shaw, and Norris (with different degrees of concern) opt for the second disjoint. As did other writers, even as they struggled to see a way out. Perhaps we are but incompletely evolved, and freedom will come eventually.

> Among the forces which sweep and play through the universe, untutored man is but a wisp in the wind. Our civilisation is still in a middle stage, scarcely beast, in that it is no longer wholly guided by instinct; scarcely human, in that it is not yet wholly guided by reason. On the tiger no responsibility rests. We see him aligned by nature with the forces of life – he is born into their keeping and without thought he is protected. We see man far removed from the lairs of the jungles, his innate instincts dulled by too near an approach to free-will, his free-will not sufficiently developed to replace his instincts and afford him perfect guidance. (Dreiser 1900, 56)

Thus Theodore Dreiser in *Sister Carrie*, a novel that tells the story of a kept woman. "In Carrie – as in how many of our worldlings do they not? – instinct and reason, desire and understanding, were at war for the mastery. She followed whither her craving led. She was as yet more drawn than she drew" (p. 57).

Is there no real escape now from the determinism, the fatalism, of evolution? A brilliant recent novel – fully informed on modern evolutionary biology – argues that there is. Not that this means that now we have an author who is entirely happy with Darwinism.

Is Biology Enough?

The novel *Enduring Love* by the British writer Ian McEwan, published in 1997, draws heavily on contemporary evolutionary ideas, particularly the sociobiological ideas of writers like Edward O. Wilson. The main theme of the story focuses on a science writer, Joe Rose, and on his being the target of an obsessed pursuit by a young man afflicted with a well-known psychological ailment – de Clérambault's syndrome – where the sufferer falls for another person in an immediate and obsessive way, convinced that the love object accepts and returns his affection. This interaction, both the pursuit and Joe's way of dealing with it (in the light of the fact that others have trouble grasping his dilemma), leads McEwan to skepticism about Darwinism, especially in the realm of morality and freedom.

The novel opens with Rose having a picnic somewhere in the English countryside with his live-in mistress, Clarissa. They see a hot-air balloon in trouble, and Joe rushes over to try to help. A number of other people are doing likewise. These other would-be helpers include a doctor in his early forties, John Logan. Unfortunately, the balloon starts to rise up with a small child trapped in the basket. All let go except for Logan, who is carried high into the air at the end of a rope. Finally, he falls several hundred feet and is killed. Particularly upsetting for Joe is the fact that John's death leaves a widow and two small children. Joe goes to visit the widow. She is desperately upset, not just by the death but because there is evidence that Logan (when he went chasing after the balloon) was not where he was supposed to be. She feares that he was about to have a picnic with some unknown woman. (A head scarf was left in Logan's car when he went running over to the balloon.)

Where does evolution fit in with all of this? At the beginning of the book, the topic of evolutionary biology is introduced explicitly, through discussions generated by Joe in the pursuit of his profession. Most particularly, there is talk of how human beings are themselves products of evolution, and of how they work together for evolutionary ends. It is explicitly acknowledged that Darwinism does not always dictate doing things exclusively for oneself, and that cooperation can be a wise and biologically driven strategy. But McEwan wants to do more than just talk about the biological ideas – he wants to show them in action, which he does through the drama at the beginning of the novel. Joe and the others are holding onto the balloon and then let go, with the exception of Logan. Why is it, Joe wonders, that they behaved as they did? They were willing to help, yes, but they were not so willing to give up their lives. Hence, with the exception of Logan, out of what one might describe as "selfishness," they let go.

I didn't know, nor have I ever discovered, who let go first. I'm not prepared to accept that it was me. But everyone claims not to have been first. What is certain is that if we had not broken ranks, our collective weight would have brought the balloon down to earth a quarter of the way down the slope a few seconds later as the gust subsided. But as I've said, there was no team, there was no plan, no agreement to be broken. No failure. So can we accept that it was right, every man for himself? Were we all happy afterwards that this was a reasonable course? We never had that comfort, for there was a deeper covenant, ancient and automatic, written in our nature. Co-operation – the basis of our earliest hunting successes, the force behind our evolving capacity for language, the glue of our social cohesion. Our misery in the aftermath was proof that we knew we had failed ourselves. But letting go was in our nature too. Selfishness is also written on our hearts. This is our mammalian conflict – what to give to the others, and what to keep for yourself. Treading that line, keeping the others in check and being kept in check by them, is what we call morality. Hanging a few feet above the Chilterns escarpment, our crew enacted morality's ancient, irresolvable dilemma: us, or me. (McEwan 1997, 14–15)

From a Darwinian perspective, what happened is perfectly understandable. The modern evolutionist thinks that we all act for selfish ends. We are back to those "selfish genes." People let go because it was in their self-interest. We are biologically programmed to act together up to a point. That is why people dashed over to the balloon. But when it becomes clear that cooperation is not the best strategy, we break the covenant. That is why people let go of the balloon. What then about Logan? Why did he behave otherwise? This dilemma is solved (in a Darwinian fashion) when it appears that Logan had a mistress. Logan was showing off in front of this mistress and so in some way was trying to promote himself. His act was directed toward his own self-advancement. The fact that it was to prove fatal was just a contingent, unfortunate consequence. Logan was rather like a peacock displaying his tail, with the intention of attracting the peahen. This is true even though sometimes the long tail proves fatal when the peacock is pursued by a predator.

Darwinian biology reigns triumphant! Although McEwan's understanding of the effects of selection is (in line with modern evolutionary thought) more sophisticated than that of Norris – he realizes that selection can lead to cooperation as well as to competition – apparently his philosophy is much the same. Darwinism shows that we are all puppets of our past – the genes, as sifted by selection, control what we do irrespective of what we think we do. But not so fast. In fact, this is the very conclusion that McEwan wants to deny. There is a final twist to the story of Dr. John Logan. At the end of the novel, thanks to the other people who were there at the site of the tragedy, we learn that Logan was entirely innocent. He had been giving a lift to two hitchhikers: a fifty-year-old professor of mathematics and his mistress. These two had good reason to keep their presence at the site anonymous. Logan therefore had not been involved in an extramarital tryst when he died. He was helping others as he was to help the child in the balloon, even unto his own death.

How do we explain this? McEwan argues that an action like this shows that biology cannot give us all of the answers. It wasn't the case that Logan's apparent altruism was bogus and that he was selfishly trying to impress his mistress. There was no mistress. Or rather, the scarf belonged to the mistress of someone else. Logan's behavior was therefore purely disinterested. He went over to help the boy in the balloon purely for noble reasons. In other words, Logan rose above his own biological selfishness. "He was a terribly brave man.... It's the kind of courage the rest of us can only dream about" (p. 230). So, in the end, we see that Logan has managed to transcend his biological ethical nature. Biology determines our actions, but by an act of free will we can rise above our biology. And of course the very title of the novel – *Enduring Love* – shows that Saint Paul (in I Corinthians 13) had a deeper insight into human nature and motivation than do the Darwinians. Humans have the potential to be morally autonomous, and even if few of us act as nobly as Logan, some of us can and do. Darwinism is inadequate if offered as a full picture of human nature.

Butler, Shaw, and Norris argued that, with Darwinism, we are stuck with biological determinism. Butler and Shaw hated this fact and wanted to replace Darwinism with a theory that had a place for human freedom. Norris recognized the fact and was resigned to it. McEwan argues that Darwinism does imply biological determinism, but he also argues that we can transcend biological determinism. No one raises the possibility that Darwinism does not necessarily imply biological determinism, and that it leaves open the possibility of some genuine form of free will. Norris does not allow that the railroad baron might have been able to control events, and McEwan does not allow that Logan might have been under the Darwinian canvas and still have acted in a noble fashion. This is a possibility that must be explored, and we shall do so in the next and final chapter, to which we now turn.

CHAPTER TWELVE

Religion

What is Darwinism? It is atheism.

"What is Darwinism?" was the question asked in the title of a book, published after the appearance of the *Origin* and authored by an American, Charles Hodge, an important Presbyterian theologian. Loud and clear came his response: "It is atheism." This would be the response of today's most ardent Darwinian scourge of the religious, Richard Dawkins. "I'm a Darwinist because I believe the only alternatives are Lamarckism or God, neither of which does the job as an explanatory principle. Life in the universe is either Darwinian or something else not yet thought of" (Brockman 1995, 85–6). If you are a Darwinian, then you ought to be an atheist, and conversely.

Not everyone thinks this way: the late nineteenth-century high Anglican clergyman Aubrey Moore wrote: "Darwinism appeared, and, under the guise of a foe, did the work of a friend." Adding: "We must frankly return to the Christian view of direct Divine agency, the immanence of Divine power from end to end, the belief in a God in Whom not only we, but all things have their being, or we must banish him altogether" (Moore 1890, 268–9). Likewise going the other way, important Darwinians have been Christians. These include Ronald Fisher (Anglican) and Theodosius Dobzhansky (Russian Orthodox) in the last century, and the paleontologist Simon Conway Morris (Anglican) in this. More common, however, are those who want to be both evolutionists and Christians, and yet cannot bring themselves to accept what they see as the harsh and undirected nature of Darwinism. Thus Keith Ward, Anglican clergyman, philosopher, and retired Regius Professor of Religion at Oxford: "On the newer, more holistic, picture, suffering and death are inevitable parts of a development that involves improvement through conflict and generation of the new. But suffering and death are not the predominating features of nature. They are rather necessary consequences or conditions of a process of emergent harmonisation which inevitably discards the old as it moves on to the new" (Ward 1996, 87).

In the same vein, Holmes Rolston III, philosopher of the environment, Presbyterian clergyman, and sometime physicist, sees life moving from simple molecules to more complex ones, and eventually into living beings. This apparently takes place through a kind of self-organization, and is combined with a kind of ratchet effect that prevents complexity from falling backward: "to have life assemble this way, there must be a sort of push-up, lock-up effect by which inorganic energy input, radiated over matter, can spontaneously happen to synthesize negentropic amino acid subunits, complex but partial protoprotein sequences, which would be degraded by entropy, except that by spiraling and folding they make themselves relatively resistant to degradation" (Rolston 1987, 111–12).

One is tempted to let Christians – believers of any kind – fend for themselves. If they want to accept Darwinism, then it is there to be accepted. If they want to reject it on religious grounds, then that is their option. But many are genuinely puzzled and concerned, and would like an answer that is not just based on the prejudice and ignorance of one side or the other. So let us end this book by looking at the relationship between Darwinism and Christianity, choosing this religion because it was that from which Darwinism emerged and against which it defined itself, but confident that our conclusions can readily be adapted and applied to other religions – especially other theistic religions (Judaism and Islam). It is traditional when talking of religious belief to make a distinction between revealed religion and natural religion, where the former refers to the area of faith and the latter to the area of reason. For the Christian, central faith beliefs include God as creator, humans as made in the image of God and Father, Jesus as Incarnation (son of God) and Saviour through the Crucifixion and Resurrection, and the possibility of eternal life. For the Christian, central reason beliefs cover the traditional proofs of God's existence, and defenses against counterattacks, especially including the problem of evil (Ruse 2001). Darwinism does not impinge directly on all of these issues. I am not sure it has a great deal to say about life after death. So on the revealed side, let us confine the discussion to creation, the status of humans, and miracles (including the Resurrection); and on the natural side, to the most significant of the proofs, the argument from design, and to the problem of evil.

Creation

Creation can mean a number of things. For the Christian, God is creator of all things, from nothing. He is not just a designer of already-existing matter. He is also in some sense the constant supporter and renewer of the creation. He is immanent – not identical to the creation, but in some way coextensive with it. Whether science has anything at all to say on these matters, it hardly seems that Darwinism does. As in the case of consciousness, for the evolutionist some things seem just to be given, and that is that. If one wants to believe (in the words of Aubrey Moore) in "the immanence of divine power from end to end," then so be it. Darwinism does not speak directly to this. It does, however, speak to another sense of creation, namely, the Genesis story of creation in six days. It speaks to it and contradicts it. The world – inorganic and organic – cannot have been created in such a fashion, nor can it have been done in five to ten thousand years (as calculated from the biblical genealogies), and there cannot have been a worldwide flood through which Noah supposedly sailed with his ark of animals, or any of the other bits and pieces, like the Tower of Babel. So if your version of Christianity makes these events central and essential, then you cannot be a Darwinian.

But one must add at once that this kind of reading of Genesis has never – at least since Saint Augustine, around 400 A.D. – been obligatory for Christians. Indeed, Augustine himself warned explicitly against this kind of literal interpretation of Scripture (McMullin 1985). On the one side, it is bound to get you into trouble, even without evolution. How, for instance, could God say "Let there be light" before the sun was created? On the other side, such a literalistic reading obscures the true message of Genesis, which is about God and humans and their relationship to Him and to His creation. About our fallen nature and our need for redemption in some way. Genesis is not false, but it is not something that can be read straight off. One must delve into the deeper, more profound meaning – about how we humans are special although flawed, and about how we have obligations to God and to His creation. None of this is in any way denied or obscured by Darwinism. Saint Augustine himself believed that God lies outside time. Hence, for Him, the thought of the creation, the act of the creation, and the product of the creation are as one. This means that in some sense God created the seeds of life from the beginning, and all else is an unfurling. This is not in itself an evolutionary picture, but there are many who would argue that it is an underlying theology that lends itself readily to an evolutionary interpretation.

Humans

But what about the nature and status of humans? If we are made in the image of God, what price evolution? And what of immortal souls, and how about our fallen nature? Evolution certainly does not deny that humans uniquely are articulate thinking beings who plan actions and who have built a culture in which they are immersed. The Darwinian approach to human nature starts with this fact. The problem is to explain it, not to deny it. There have always been theological problems about the nature of the soul and its relationship to consciousness (Brown, Murphy, and Malony 1998). On the one hand, it is obviously connected to consciousness – consciousness is what we are talking about when we say that humans are made in the image of God – on the other hand, it cannot be identical to consciousness, at least not without denying souls (or fully functioning souls) to children and idiots and the like. What one can say is that even if Darwinism does not explain the nature and existence of consciousness, it does care about the reasons why it might have evolved as it has in humans. To say more than this is to take you into theological matters. If (in line with the thinking of the late Pope) you want to argue that souls are something non-natural, then this demands intervention of a kind that is simply beyond the scope and ken of science (John Paul II 1997). The Darwinian feels a little uncomfortable about one generation of proto-humans not having souls and the next generation having souls, but that is an issue for the believer and not for the scientist qua scientist.

Surely Christianity cannot accept that humans just arrived by chance? Obviously this question is much bound up with questions of progress, and we have already discussed these extensively. If you are demanding progress, you are probably not going to get it. We have seen no reason to think that Darwinism produces humans as an inevitable highest product of the evolutionary process. However, if you are prepared to impute worth rather than trying to find it existing independently in nature – if you yourself judge human-like qualities to be those that you want – and if you are simply asking if human-like beings are likely to evolve, then perhaps your quest is not so unreasonable. If Conway Morris's (2003) arguments hold, then perhaps human-like creatures were indeed bound to emerge, even though he certainly has not proven (through biology) that humans are superior. This is enough for the Christian.

What of original sin and our fallen nature? If one insists that souls came into being only at one miraculous moment, the believer is committed then to the belief that real sin came only after this event, perhaps through some act of disobedience or immorality. If one insists that this sin must have been the defining moment for all humans, and that it is only the descendents of this sinner who are thus tainted with original sin, then presumably this can be fitted somehow with Darwinism, given that it is now believed that there were individual proto-humans (or real humans) from whom we are all descended. But all of this is starting to seem rather strained – we may all be descended from one human, but that does not mean that we are all descended from only one human or even that there is only one human from whom we are all descended (Ayala 1995). Was there one person who was a sinner who had a mate who was not, or were they both sinners but their parents and siblings were not?

Better to pull back somewhat and let Darwinism do some of the work. Try to relate sin itself to our biological nature. Despite the author's personal dissatisfaction with pure Darwinism, *Enduring Love* shows us how this can be done. Humans are not all bad. We cooperate and work together. Humans are not all good. We are selfish and serve our own ends rather than the needs of others. This is our nature, and it is precisely the nature that we find supposed at the heart of Christianity. We are born this way, and it is no chance – it is part of our heritage. Not a literal Adam who ate an apple, but proto-humans forged by natural selection who survived and reproduced by being both cooperative and selfish. And we are not going to change, whatever our intentions. It is as silly to think that we are all going to do good because of some internal motivation or external exhortation, as it is to think that always and in every way we are going to be mean and selfish. In this respect, there is a perfect consilience between the Darwinian human and the Christian human.

Miracles

The Christian believes that humans can be saved by the death of Christ on the Cross and his subsequent Resurrection. The Christian believes that this salvation will come after death – or at the Second Coming. On these matters, Darwinism is silent. But what about the Resurrection, and indeed all of the other miracles surrounding the Christian story? What about the feeding of the five thousand and Christ walking on water? What about the subsequent miracles of the Apostles, or the miracles supposedly still occurring – the marks the Catholic Church seeks for admission to sainthood? Some supposed miracles are by their very nature put beyond the bounds of science. The appearance of souls – whether it occurred just once and then was transmitted, or occurs for each individual – is something about which science can say nothing. The same is true of transubstantiation, the miracle that occurs in the Catholic mass, when the water and wine is turned into the body and blood of Christ. No amount of microscopic examination of the host is going to reveal red corpuscles. It is just not that sort of miracle.

But what about rising from the dead and turning water into wine? Darwinism is a scientific theory, and scientific theories exclude miracles – that is what they are all about, working through laws. There are two (traditional) approaches one can take (Swinburne 1970; Mullin 1996). The first, stemming from Saint Augustine, interprets miracles for their spiritual meaning rather than seeing them as violations of law. Thus, to take the miracle at Cana (water into wine), the real miracle was not some jiggery pokery that was shortcutting the fermenting process, but the fact that the man throwing the party supplied his guests fully, even bringing out his very best wine when they had no reason to expect it. Jesus filled him with such love that he went against his usual nature. (Can any one of my readers deny having brought out the cheap plonk when the guests were well lubricated?) Even the miracle of the Resurrection can be treated this way. The real miracle was not some reversal of life-death processes, but that, on the third day, the disciples who were downcast and lonely suddenly felt a great lift and that life was meaningful for them – that Jesus had left a message and example that they wanted to promulgate. If some psychologist explains this in terms of mass hysteria or whatever, so be it. There will always be a natural explanation. This leaves the meaning of the event untouched.

The second approach, stemming from Saint Thomas, thinks that events outside the course of nature did occur – whether these broke laws or stepped around them. Here one simply has to invoke the traditional distinction (raised by Ernan McMullin [1991] in reply to Plantinga) between the order of nature and the order of grace. Normally, life does work by law, but in special circumstances – as when bringing salvation to humankind – God steps in and works directly. He holds the world in his hands at all times, and thus for Him to do this is no violation of His nature or powers. Obviously there is going to be more tension for the Darwinian if this approach is taken – especially if the believer tries to justify the exceptions rather than simply take them on faith – but the believer can point out that miracles are really miraculous only if they are rare. The success of science underlines this point. If water turned into wine on a regular basis, then Jesus' actions at the wedding would call for no special comment.

Design

Let us turn now from revealed religion to natural religion. Darwinism does not impinge directly on every aspect. The ontological argument, trying to derive God's existence from His nature, does not connect with biology at all. But the argument from design, the teleological argument, does. So let us focus on that (Ruse 2003).

The tale is simple. Organisms show design-like features: the hand, the eye, the leaf, the fin, the funny plates on the back of stegosaurus. These "adaptations" function, they work, for the benefit of their possessors. They seem far too complex to have come about by chance. In the real world, things break down and go wrong, rather than build up into working units. So there must be an explanation. And the only reasonable explanation is that there is a designer, an intelligence behind adaptations. The eye is like a telescope. Artifacts have designers – us humans. Therefore, adaptations must have designers – and if not us humans, then God is the only possibility. For the great philosophers of the past, it was the Christian God: the world "by the great beauty of all things visible, proclaims by a kind of silent testimony of its own both that it has been created, and also that it could not have been made other than by a God ineffable and invisible in greatness, and ineffable and invisible in beauty" (Augustine 1998, 452–3). Then along came Darwin, and no longer was God the only game in town. Natural selection explains adaptation – the design-like features of organisms were forged in the struggle for existence, as their possessors battled to live and to reproduce. There is no need of conscious intention. Remember what Richard Dawkins (1986) has said: after (and only after) the *Origin* is it possible to be an intellectually fulfilled atheist. Natural theology has collapsed.

Intelligent Design

One obvious response is to deny that natural selection can explain adaptations fully. Hence there is still a place (even a necessity) for a designer. This is the position (mentioned in an earlier chapter) of a group of (mainly) American evangelical Christians, the so-called Intelligent Design theorists (Pennock 1998; Ruse 2005b). Their argument is that the organic world, particularly at the microscopic level, shows such intricate functioning complexity – what they call "irreducible complexity" – that it could not have been produced by the blind process of natural selection. There must be something else, and (in the words of their English sympathizer Anthony Flew) "there must have been some intelligence." Defining irreducible complexity as "a single system composed of several well-matched, interacting parts that contribute to the basic function, wherein the removal of any one of the parts causes the system to effectively cease functioning" (Behe 1996, 39), the most important proponent of Intelligent Design, the Lehigh University biochemist Michael Behe, writes that any "irreducibly complex biological system, if there is such a thing, would be a powerful challenge to Darwinian evolution. Since natural selection can only choose systems that are already working, then if a biological system cannot be produced gradually it would have to arise as an integrated unit, in one fell swoop, for natural selection to have anything to act on" (ibid.).

As an example of irreducible complexity that is human-made, Behe instances a mousetrap. It is made of several parts – spring, snapper, base, and the like – and it will not work until and unless it has been assembled into a complex whole. Take out one part, and it fails to function. It cannot have evolved gradually but must have been put together by a person at one point in time. Likewise in nature. An example is furnished by bacteria, specifically those that use a flagellum (a kind of whip-like strand), powered by a sort of rotary motor, to propel themselves along. Everything is highly complex, and nothing will function until everything is in place. To take an example: the "flagellin," the external filament of the flagellum, is a single protein. It makes a kind of paddle surface that contacts the liquid during swimming. Near the surface of the cell, one finds a thickening – just as needed, so that the filament can be connected to the rotor drive. In turn, we need a connector, something known as a "hook protein." The filament has no motor. It has to be somewhere else. "Experiments have demonstrated that it is located at the base of the flagellum, where electron microscopy shows several ring structures occur" (p. 70). The conclusion is simple and obvious. This whole system is far too complex to have come into being in a gradual fashion. It had to be formed in one step, and such a process must involve some sort of designing cause.

This whole line of thought has been much publicized and equally much criticized, and there is no need to dwell on it here (Doolittle 1997; Miller 1999). No one has ever seen such a one-step (presumably miraculous) process. Failure of observation does not make something impossible or nonexistent, but it does rather suppose that the creative acts bringing on irreducible complexity occurred in the past, perhaps long in the past, and that the new features had to wait to be brought into action. Which in turn raises the question of why they did not degrade through mutation and selection before they were used. It is well known that if a feature (like the eyes) is not being used, then selection acts to remove or downgrade it. Complex functions are high-maintenance; if they aren't used, then they are eliminated. Moreover, Behe shows great ignorance of the way in which Darwinian evolution works. Apart from the fact that his example of a mousetrap is not necessarily an all-or-nothing artifact – critics have shown much ingenuity in making functioning traps with fewer parts – complex organic systems do not necessarily have to come into being in one step or not at all. Generally, such systems are cobbled together from parts that are already functioning in the system and then are turned to other uses.

Take the Krebs cycle, a highly complex process with many steps, used by the cell to provide energy. It did not just spring into being. It was a "bricolage," built bit by bit from other pieces.

> The Krebs cycle was built through the process that Jacob (1977) called "evolution by molecular tinkering," stating that evolution does not produce novelties from scratch: It works on what already exists. The most novel result of our analysis is seeing how, with minimal new material, evolution created the most important pathway of metabolism, achieving the best chemically possible design. In this case, a chemical engineer who was looking for the best design of the process

could not have found a better design than the cycle which works in living cells. (Melendez-Hevia, Waddell, and Cascante 1996, 302)

Note also how Intelligent Design digs a horrendous hole into which fall the designer's good intentions. Many vile afflictions are caused by minor changes at the molecular level. The effects multiply, bringing on lifelong pain and suffering. If the designer is around to make the very complex, why doesn't he take a little time to repair the simple but broken? Either he cannot, in which case one wonders just how powerful he really is and if he truly has designed the very complex; or he does not, in which case one wonders about his intentions toward the world of life, including humans. Either way, the designer seems not to be something that can be identified with the Christian God, which is the underlying aim of the Intelligent Design theorists.

There is more that could be said about Intelligent Design, for instance, about its apparent inability to lead to any new scientific predictions – indeed, it flunks just about all of the epistemic criteria for good science. But the point is made. It is not going to resurrect the argument from design.

The Problem of Evil

It is possible to be an intellectually fulfilled atheist. Is this the strongest conclusion that can be drawn? Perhaps after Darwin one should be an intellectually fulfilled atheist. Of course, there has to be more to the case than this. It is one thing to say that something is not necessary. It is another to say that it is impossible. Critics like Richard Dawkins (1995, 2003b) are quite open about their grounds of attack. It is that Darwinism, based as it is on a struggle in which anything that will help is preserved, draws attention to pain and suffering and especially to the problem of evil. It makes unreasonable belief in the Christian God, loving and powerful. Darwin himself made this case, in a letter written just after the publication of the *Origin* to his American friend and supporter Asa Gray. He wrote:

With respect to the theological view of the question; this is always painful to me. – I am bewildered. – I had no intention to write atheistically. But I own that I cannot see, as plainly as others do, & as I sh^{d}. wish to do, evidence of design & beneficence on all sides of us. There seems to me too much misery in the world. I cannot persuade myself that a beneficent & omnipotent God would have designedly created the Ichneumonidae with the express intention of their feeding within the living bodies of caterpillars, or that a cat should play with mice. Not believing this, I see no necessity in the belief that the eye was expressly designed. (letter to Asa Gray, May 22, 1860)

As it happens, at the time of this writing, Darwin still believed in a deistic God who works through unbroken law. But many have taken this kind of point to the limit. "The universe we observe has precisely the properties we should expect if there is, at bottom, no design, no purpose, no evil and no good, nothing but blind, pitiless indifference" (Dawkins 1995, 131–3).

What can be said in reply? Most obviously, even if Darwinism exacerbates the problem, it does not create it. Christians have long wrestled with the problem of evil. And they have counterarguments. These may not satisfy everyone, but that is not quite the point. If Christians find their responses adequate, then unless Darwinism shows (on its own merits) that there are further difficulties, we can leave matters at that. The Christian has no reason to focus on Darwinism in particular as a cause for discontent. What then are the Christian's responses? It is customary to distinguish between moral evil – the evil caused by humans (Auschwitz) – and physical or natural evil – the evil caused by natural phenomena (earthquakes and diseases) (Reichenbach 1982). Moral evil is answered in terms of free will and its desirability. Natural evil is answered in terms of inevitability. Let us take these answers in turn.

Free Will

The free-will argument against moral evil is simply that it is a better world if we humans have free will than if we are determined. Even though we can and do perform evil, better this than that we all are robots with no choice. There are a number of standard objections to this argument – for instance, why are we not free and yet doing good all of the time? – but let us concentrate purely on those points that impinge on Darwinism. Here we must at last tackle the free-will issue (Ruse 1987). There are two levels at which we can address this question. First, one can ask if any scientific theory threatens free will. Since Darwinism is a scientific theory, will Darwinism considered just as a scientific theory threaten free will? Many think that it must, because (as noted earlier) scientific theories presuppose a law-bound world, and this seems to preclude freedom. A falling rock has no freedom. Why then should a law-bound human have freedom?

There is a standard reply to this question, given most fully by David Hume (1978). Freedom is not the contradictory of being law-bound. If someone is not bound by law, they are not free. They are crazy. Suppose I take my clothes off in the middle of downtown Tallahassee, my home. Suppose I do so because for some reason the laws of nature have been lifted. If anything is a random, uncaused act, this is. So why blame me? If I took my clothes off to make a protest, like the Dukabors do, you might disapprove, but you would not think me mad. You would see that my actions had been brought on by previous states – for instance, a desire to protest America's foreign policy, or perhaps an unresolved childhood urge to exhibitionism – and it is precisely because you could fit me into such a framework that you would think me responsible. As Hume argued, the disjunction should be between freedom and restraint, not freedom and law. If I strip because I have been hypnotized or because I am being threatened – some enemy wants to humiliate me in public – you should exonerate me because then I am truly not a free agent.

What about Darwinism in particular and free will? We return to the charge that Darwinism applied to human social behavior makes us all "genetically determined." We are marionettes, controlled from above by the double helix. We are programmed by our biology. We have no freedom, good or ill, because our genes made us do it. Hitler is not to blame. He just had a lousy genotype, and it is natural selection that put that in place. Blame the process, not us. Likewise, of course, Mother Teresa is not to be praised. She drew a good genotype. This argument did not stand up before, and it does not stand up now. Some organisms are surely genetically determined – ants, for instance. They are preprogrammed by the genes, as produced by selection (Dennett 1984). But we humans are not ants – we are not genetically determined in this way. Our evolution has given us the power to make decisions when faced with choices, and to revise and rework them when things go wrong.

In the language of evolutionists, ants were produced by "r-selection." They produce lots of offspring, and when something goes wrong they can afford to lose them, because there are more. We humans are "K-selected." We produce just a few offspring, and we cannot afford to lose them when things go wrong. Hence, we have the ability to make decisions in order to avoid obstacles. That is why we have big brains. In a way, ants are like cheap rockets – many are produced, and they cannot change course when once fired. We humans, by contrast, are like expensive rockets – just a few are produced, but we can change course even in midflight, if the target changes direction or speed or whatever. The expensive rocket has flexibility – a dimension of freedom – not possessed by the cheap rocket. Both kinds of rockets are covered by laws, and so are ants, wasps, and humans. We have freedom over and above genetic determinism, and this freedom was put in place by – not despite – natural selection.

Is this freedom enough to explain the actions of someone like John Logan in *Enduring Love*? If you insist that Logan somehow broke from his biology, then obviously it is not. This kind of freedom exists because of the biology, not despite it. Logan, Joe Rose, and the others had the choice to hang on or to let go. No one was constraining or forcing them. They were not programmed to do what they did. If they had been, Rose would have felt no guilt, and Logan would have merited no praise. The question is more whether an evolutionary approach can explain or understand such a disinterested action as that of Logan. That he performed an action that led to his death is perhaps in itself not a major issue. The whole point about the evolutionary approach is that morality is a quick-and-dirty solution. One makes a decision and gets on with it, rather than taking time to calculate. Sometimes quick-and-dirty solutions have costs – generally not costs as great as Logan's, but then generally we are not hanging on to a hot-air balloon with a kid alone in the basket. The major issue really is why Logan, uniquely, hung on. In part, obviously, because although it was at a moment of great threat, he did not realize how great a threat. If Logan had realized that his hanging on was worth nothing and that he would be killed, he would have let go.

But why Logan and not the others? Can this be explained by biology, or do we need to bring in something else? Let us assume that Logan was not just a slow thinker or whatever, but that he was genuinely a nicer human being. Two points are relevant. One is that biology expects variation, so if Logan was nicer than the others, so be it. The second is that this is really not a question of biology versus something else, where that something else is probably culture (which includes moral training, religious heritage, and so forth). To say again what has been said many times in this book, human beings are biological beings immersed in the culture that they have created. Note that Logan was the doctor and not the science writer. Medical culture demands giving and sacrifice – doctors are expected to go to a plague center when the rest of us are warned off – and so the fact that it was Logan who made the extra effort on the balloon is expected. The real point is that it is not a question of morality versus the genes, or even morality through culture versus the genes. It is more a matter of morality as produced by culture informed by the genes that made Logan and the others act as they did. They were free beings. One did one thing. The others did another thing. That the other thing was not as noble is expected both by Darwinism and Christianity.

Not Doing the Impossible

Finally, turn to the issue of natural evil. Darwinism does highlight pain and suffering. But is this a counter to Christianity? The traditional refuting argument is one that is usually associated with the great German philosopher Leibniz. Disagreeing with Descartes about the powers of supernatural beings, he pointed out that being all-powerful has never implied the ability to do the impossible. God cannot make $2 + 2 = 5$. No more can God, having decided to create through law (and surely an Augustinian would think that there may be good theological reasons for this), make physical evil disappear. Indeed, it may well be that physical evil simply comes as part of a package deal. "For example, what would it entail to alter the natural laws regarding digestion, so that arsenic or other poisons would not negatively affect my constitution? Would not either arsenic or my own physiological composition or both have to be altered such that they would, in effect, be different from the present objects which we now call arsenic or human digestive organs?" (Reichenbach 1976, 185)

Paradoxically and somewhat amusingly, Richard Dawkins (1983) aids this line of argument. He has long maintained that the only way in which complex adaptation could be produced by law is through natural selection. He argues that alternative mechanisms (notably Lamarckism) that produce adaptation are false, and that alternative mechanisms (notably evolution by jumps, or saltationism) that do not produce adaptation are inadequate. "If a life-form displays adaptive complexity, it must possess an evolutionary mechanism capable of generating adaptive complexity. However diverse evolutionary mechanisms may be, if there is no other generalization that can be made about life all around the Universe, I am betting that it will always be recognizable as Darwinian life" (p. 423). In short, if God was to create through law, then it had to be through

Darwinian law. There was no other choice. (This of course is not to say that, knowing the subsequent pain, God was right to create at all, but that is another matter and none of Darwinism's business.)

Conclusion

I am not trying to defend Christianity. Rather, the question is whether the Christian has good grounds for picking out Darwinism as something especially threatening, to be rejected if one's faith is to be secured. My answer is that it is not. To conclude, let us turn the question around for a moment. Does Darwinism have something to offer the Christian? Something positive? Darwinism has undermined the argument from design, and it is wise to let all of natural theology go that way. Too often a new scientific discovery is held up as proof of the truth of Christianity. The Big Bang is a favorite example; more recently, the claim that (on the basis of shared mitochondria) all humans are descended from one woman who lived about 150,000 years ago: "mitochondrial Eve." The problem is that the science rarely proves exactly what the believer wants – mitochondrial Eve was not the only female or even necessarily the only female from whom we are all descended – and when the science proves fallible, the believer is left looking silly. Better by far not to get into this business in the first place.

This does not mean that nature – as revealed by Darwinism – is irrelevant to faith. Obviously, the world counts. One uses the world to illuminate and to flesh out faith, rather than to prove it. One has what has been called a "theology of nature" rather than a "natural theology" (Pannenberg 1993). As so often, John Henry Newman had it right. To a correspondent about his seminal philosophical work *A Grammar of Assent*, Newman wrote:

> I have not insisted on the argument from *design*, because I am writing for the 19th century, by which, as represented by its philosophers, design is not admitted as proved. And to tell the truth, though I should not wish to preach on the subject, for 40 years I have been unable to see the logical force of the argument myself. I believe in design because I believe in God; not in a God because I see design. (Newman 1973, 97)

Continuing: "Design teaches me power, skill and goodness – not sanctity, not mercy, not a future judgment, which three are of the essence of religion."

One final point. This has been a book explaining Darwinism and defending it against its detractors. It is a wonderful theory. If ever proof were needed that we are more than grubby little primates, it is our finding and articulating the theory of evolution through natural selection. If ever the believer needed reassurance that, through our powers of understanding, we are made in the image of God, our grasp of life's evolutionary principles provides it. The Christian, however, has always stressed that despite our great powers, we are limited. We see as "through a glass darkly." Ultimately, all is shrouded in mystery, not to be revealed in this lifetime. Is this not the ultimate betrayal for the Darwinian? Not at all. The Darwinian concurs! The Darwinian has no time for science stoppers or for duds like Intelligent Design. However, Darwinism at base is a scientific theory – you can make metaphysics out of it, but that is not true Darwinism, or at least it is an optional extension. It cannot – it should not – say everything. This does not mean that the way is open for foggy-minded mysticism. But Darwinism realizes that it – all human knowledge – has limits. We evolved with adaptations to get out of the jungle and up on our hind legs. We evolved to survive and reproduce, to get food and shelter and mates and to raise our children. We did not evolve to get insights into life's ultimate realities. What we do find is a bonus, and we expect that bonus to run out at some point.

As Richard Dawkins has truly said: "Modern physics teaches us that there is more to truth than meets the eye; or than meets the all too limited human mind, evolved as it was to cope with medium-sized objects moving at medium speeds through medium distances in Africa" (Dawkins 2003b, 19). Perhaps there is God on the other side. Dawkins thinks not. I do not know. The Christian thinks so. Darwinism's inability to give an answer – or rather, its pointing to our inability to give a generally accepted answer – is not a weakness. It is strength. Beware of anything that answers everything. It usually ends by answering nothing. And that is certainly not true of Darwinism.

References

Agassiz, E. C., editor. 1885. *Louis Agassiz: His Life and Correspondence.* Boston: Houghton Mifflin.

Alcock, J. 2001. *The Triumph of Sociobiology.* New York: Oxford University Press.

Alvarez, L. W., W. Alvarez, F. Asaro, and H. V. Michel. 1980. Extraterrestrial cause for the Cretaceous-Tertiary extinction. *Science* 208: 1095–108.

Anton, S. C., W. R. Leonard, and M. L. Robertson. 2002. An ecomorphological model of the initial hominid dispersal from Africa. *Journal of Human Evolution* 43: 773–85.

Aquinas, T. 1952. *Summa Theologica, I.* London: Burns, Oates and Washbourne.

Aristotle. 1984. De Anima. In *The Complete Works of Aristotle,* edited by J. Barnes. Princeton, N.J.: Princeton University Press.

Augustine. [413–26] 1998. *The City of God against the Pagans*, edited and translated by R. W. Dyson. Cambridge: Cambridge University Press.

Ayala, F. J. 1974. Introduction. In *Studies in the Philosophy of Biology*, edited by F. J. Ayala, and T. Dobzhansky, vii, xvi. Berkeley: University of California Press.

Ayala, F. J. 1985. The theory of evolution: recent successes and challenges. In *Evolution and Creation*, edited by E. McMullin, 59–90. Notre Dame, Ind.: University of Notre Dame Press.

Ayala, F. J. 1995. The myth of Eve: molecular biology and human origins. *Science* 270: 1930–36.

Ayala, F. J. 1998. Human nature: one evolutionist's view. In *Whatever Happened to the Soul? Scientific and Theological Portraits of Human Nature*, edited by W. S. Brown, N. Murphy, and H. N. Malony, 31–48. Minneapolis: Fortress Press.

Bakker, R. T. 1983. The deer flees, the wolf pursues: incongruencies in predator-prey coevolution. In *Coevolution*, edited by D. J. Futuyma and M. Slatkin. Sunderland, Mass.: Sinauer.

Bannister, R. 1979. *Social Darwinism: Science and Myth in Anglo-American Social Thought*. Philadelphia: Temple University Press.

Barkow, J. H., L. Cosmides, and J. Tooby, editors. 1991. *The Adapted Mind: Evolutionary Psychology and the Generation of Culture.* New York: Oxford University Press.

Barrett, P. H., P. J. Gautrey, S. Herbert, D. Kohn, and S. Smith, editors. 1987. *Charles Darwin's Notebooks, 1836–1844.* Ithaca, N.Y.: Cornell University Press.

Barton, N. H., and M. Turelli. 2004. Effects of allele frequency changes on variance components under a general model of epistasis. *Evolution* 58: 2111–32.

Bates, H. W. [1862] 1977. Contributions to an insect fauna of the Amazon Valley. In *Collected Papers of Charles Darwin*, edited by P. H. Barrett, 87–92. Chicago: University of Chicago Press.

Beatty, J. 1978. "Evolution and the Semantic View of Theories." Unpublished Ph.D. thesis, Indiana University.

Beatty, J. 1981. What's wrong with the received view of evolutionary theory? *PSA 1980* 2: 397–426. East Lansing, Mich.: Philosophy of Science Association.

Beatty, J. 1985. Speaking of species: Darwin's strategy. In *The Darwinian Heritage*, edited by David Kohn, 265–81. Princeton, N.J.: Princeton University Press.

Behe, M. 1996. *Darwin's Black Box: The Biochemical Challenge to Evolution*. New York: Free Press.

Benton, M. J. 1987. Progress and competition in macroevolution. *Biological Reviews* 62: 305–38.

Benton, M. J. 1990. *Vertebrate Paleontology*. London: Unwin Hyman.

Benton, M. J. 1997. Models for the diversification of life. *TREE* 12: 490–5.

Benton, M. J. 2007. Paleontology. In *Harvard Companion to Evolution*, edited by M. Ruse and J. Travis. Cambridge, Mass.: Harvard University Press.

Benton, M. J., and G. W. Storrs. 1994. Testing the quality of the fossil record: paleontological knowledge is improving. *Geology* 22: 111–14.

Bergson, H. 1907. *L'évolution créatrice*. Paris: Alcan.

Bergson, H. 1911. *Creative Evolution*. New York: Holt.

Berry, R. J. 1986. Genetics of insular populations of mammals, with particular reference to differentiation and founder effects in British small mammals. *Biological Journal of the Linnaean Society* 28: 205–30.

Blackmore, S. 2000. *The Meme Machine*. Oxford: Oxford University Press.

Bowler, P. J. 1976. *Fossils and Progress*. New York: Science History Publications.

Bowler, P. J. 1986. *Theories of Human Evolution*. Baltimore: Johns Hopkins University Press.

Bowler, P. J. 1988. *The non-Darwinian Revolution: Reinterpreting a Historical Myth*. Baltimore: Johns Hopkins University Press.

Bowler, P. J. 2005. Revisiting the eclipse of Darwinism. *Journal of the History of Biology* 38: 19–32.

Bowring, S. A., and T. Housh. 1995. The earth's early evolution. *Science* 269: 1535–40.

Boyle, R. [1688] 1966. A disquisition about the final causes of natural things. In *The Works of Robert Boyle*, edited by T. Birch, vol. 5: 392–444. Hildesheim: Georg Olms.

Brack, A., editor. 1999. *The Molecular Origins of Life: Assembling Pieces of the Puzzle*. Cambridge: Cambridge University Press.

Brackman, A. C. 1980. *A Delicate Arrangement: The Strange Case of Charles Darwin and Alfred Russel Wallace*. New York: Times Books.

Brandon, R. N., and R. M. Burian, editors. 1984. *Genes, Organisms, Populations: Controversies over the Units of Selection*. Cambridge, Mass.: MIT Press.

Brockman, J. 1995. *The Third Culture: Beyond the Scientific Revolution*. New York: Simon and Schuster.

Brocks, J. J., G. A. Logan, R. Buick, and R. E. Summons. 1999. Archean molecular fossils and the early rise of eukaryotes. *Science* 285: 1033–6.

Brown, P., T. Sutikna, M. J. Morewood, R. P. Soejono, E. Jatmiko, E. Wayhu Saptomo, and Rokus Awe Due. 2004. A new small-bodied hominin from the Late Pleistocene of Flores, Indonesia. *Nature* 431: 1055–61.

Brown, W. S., N. Murphy, and H. N. Malony, editors. 1998. *Whatever Happened to the Soul? Scientific and Theological Portraits of Human Nature*. Minneapolis: Fortress Press.

Browne, J. 1995. *Charles Darwin: Voyaging. Volume I of a Biography*. New York: Knopf.

Browne, J. 2002. *Charles Darwin: The Power of Place. Volume II of a Biography*. New York: Knopf.

Brunet, M., F. Guy, D. Pilbeam, and others. 2002. A new hominid from the Upper Miocene of Chad Central Africa. *Nature* 418: 145–51.

Burchfield, J. D. 1975. *Lord Kelvin and the Age of the Earth*. New York: Science History Publications.

Buri, P. 1956. Gene frequencies in small populations of mutant *Drosophila*. *Evolution* 10: 367–402.

Bury, J. B. [1920] 1924. *The Idea of Progress; An Inquiry into Its Origin and Growth*. London: Macmillan.

Butler, S. 1872. *Erewhon, or Over the Range*. London: Trubner.

Butler, S. 1879. *Evolution, Old and New*. London: Hardwicke and Bogue.

Butler, S. 1901. *Erewhon Revisited*. London: Grant Richards.

Butler, S. 1903. *The Way of All Flesh*. London: Grant Richards.

Butterworth, B. 1999. *What Counts: How Every Brain is Hardwired for Math*. New York: Simon and Schuster.

Caccone, A. E., E. N. Moriyama, J. M. Gleason, L. Nigro, and J. R. Powell. 1996. A molecular phylogeny for the *Drosophila melanogaster* subgroup. *Molecular Biology and Evolution* 13: 1224–32.

Cairns-Smith, A. G. 1986. *Clay Minerals and the Origin of Life*. Cambridge: Cambridge University Press.

Campbell, D. T. 1974. Evolutionary epistemology. In *The Philosophy of Karl Popper*, edited by P. A. Schilpp, vol. 1: 413–63. LaSalle, Ill.: Open Court.

Cann, R. L., and A. C. Wilson. 2003. The recent African genesis of humans. *Scientific American* 13, no. 2: 54–61.

Carroll, R. L. 1997. *Patterns and Processes of Vertebrate Evolution*. Cambridge: Cambridge University Press.

Carroll, S. B., J. K. Grenier, and S. D. Weatherbee. 2001. *From DNA to Diversity: Molecular Genetics and the Evolution of Animal Design*. Oxford: Blackwell.

Cavalli-Sforza, L. L., and W. F. Bodmer. 1971. *The Genetics of Human Populations*. San Francisco: W. H. Freeman.

Chambers, G. K. 1988. The *Drosophila* alcohol dehydrogenase gene-enzyme system. In *Advances in Genetics*, edited by E. W. Caspari and J. G. Scandalois, vol. 25: 39–107. New York: Academic Press.

Chambers, R. 1846. *Vestiges of the Natural History of Creation*, 5th ed. London: J. Churchill.

Charlesworth, B., R. Lande, and M. Slatkin. 1982. A Neo-Darwinian commentary on macroevolution. *Evolution* 36: 474–98.

Chesterton, G. K. 1986. *The Complete Father Brown*. Harmondsworth, Mddx.: Penguin.

Chomsky, N. 1957. *Syntactic Structures*. The Hague: Mouton.

Chomsky, N. 1966. *Cartesian Linguistics*. New York: Harper and Row.

Clutton-Brock, T. H., F. E. Guinness, and S. D. Albon. 1982. *Red Deer: Behaviour and Ecology of the Two Sexes*. Chicago: University of Chicago Press.

Committee on Strategies for the Management of Pesticide Resistant Pest Populations, editors. 1986. *Pesticide Resistance: Strategies and Tactics for Management*. Washington, D.C.: National Academy Press.

Conway Morris, S. 1998. *The Crucible of Creation: The Burgess Shale and the Rise of Animals*. Oxford: Oxford University Press.

Conway Morris, S. 2003. *Life's Solution: Inevitable Humans in a Lonely Universe*. Cambridge: Cambridge University Press.

Corsi, P. 2005. Before Darwin: Transformist concepts in European natural history. *Journal of the History of Biology* 38: 67–83.

Cosmides, L. 1989. The logic of social exchange: has natural selection shaped how humans reason? Studies with the Wason selection task. *Cognition* 31:187–276.

Cox, C. D., and P. D. Moore. 1993. *Biogeography: An Ecological and Evolutionary Approach*. Oxford: Blackwell.

Coyne, J. A. 1998. Not black and white. *Nature* 396: 35–6.

Coyne, J. A., N. H. Barton, and M. Turelli. 1997. Perspective: a critique of Sewall Wright's shifting balance theory of evolution. *Evolution* 51, no. 3: 643–71.

Coyne, J. A., and H. A. Orr. 2004. *Speciation*. Sunderland, Mass.: Sinauer.

Cuvier, G. 1813. *Essay on the Theory of the Earth*, translated by Robert Kerr. Edinburgh: W. Blackwood.

Cuvier, G. 1817. *Le règne animal distribué d'aprés son organisation, pour servir de base à l'histoire naturelle des animaux et d'introduction à l'anatomie comparée*. Paris.

Daly, M., and M. Wilson. 1988. *Homicide*. New York: De Gruyter.

Daly, M., and M. Wilson. 1998. *The Truth about Cinderella: A Darwinian View of Parental Love*. London: Weidenfeld and Nicholson.

Darlington, C. D. 1959. *Darwin's Place in History*. Oxford: Blackwell.

Darwin, C. 1859. *On the Origin of Species*. London: John Murray.

Darwin, C. 1871. *The Descent of Man*. London: John Murray.

Darwin, C. 1959. *The Origin of Species by Charles Darwin: A Variorum Text*, edited by M. Peckham. Philadelphia: University of Pennsylvania Press.

Darwin, C., and A. R. Wallace. 1958. *Evolution by Natural Selection*. Foreword by Gavin de Beer. Cambridge: Cambridge University Press.

Darwin, E. 1794–96. *Zoonomia; or, The Laws of Organic Life*. London: J. Johnson.

Darwin, E. 1803. *The Temple of Nature*. London: J. Johnson.

Darwin, F. 1887. *The Life and Letters of Charles Darwin, Including an Autobiographical Chapter*. London: Murray.

Davies, N. B. 1992. *Dunnock Behaviour and Social Evolution*. Oxford: Oxford University Press.

Davies, N. B., and M. de L. Brooke. 1988. Cuckoos versus reed warblers: adaptations and counter-adaptations. *Animal Behaviour* 36: 262–84.

Davies, P. 1999. *The Fifth Miracle: The Search for the Origin and Meaning of Life*. New York: Simon and Schuster.

Dawkins, R. 1976. *The Selfish Gene*. Oxford: Oxford University Press.

Dawkins, R. 1982. *The Extended Phenotype: The Gene as the Unit of Selection*. Oxford: W. H. Freeman.

Dawkins, R. 1983. Universal Darwinism. In *Molecules to Men*, edited by D. S. Bendall. Cambridge: Cambridge University Press.

Dawkins, R. 1986. *The Blind Watchmaker*. New York: Norton.

Dawkins, R. 1995. *A River Out of Eden*. New York: Basic Books.

Dawkins, R. 2003. *A Devil's Chaplain: Reflections on Hope, Lies, Science and Love*. Boston and New York: Houghton Mifflin.

Dawkins, R. 2004. *The Ancestor's Tale*. London: Weidenfeld and Nicholson.

De Waal, F. 1982. *Chimpanzee Politics: Power and Sex among Apes*. London: Cape.

De Waal, F. 1996. *Good Natured: The Origins of Right and Wrong in Humans and Other Animals*. Cambridge, Mass.: Harvard University Press.

Deacon, T. W. 1997. *The Symbolic Species: The Co-Evolution of Language and the Brain*. New York: Norton.

Dembski, W. A. 1998a. *The Design Inference: Eliminating Chance through Small Probabilities*. Cambridge: Cambridge University Press.

Dembski, W. A., editor. 1998b. *Mere Creation: Science, Faith and Intelligent Design*. Downers Grove, Ill.: Intervarsity Press.

Dennett, D. C. 1984. *Elbow Room*. Cambridge, Mass.: MIT Press.

Dennett, D. C. 1995. *Darwin's Dangerous Idea*. New York: Simon and Schuster.

Depew, D. J., and B. H. Weber. 1985. Innovation and tradition in evolutionary theory: an interpretive afterword. In *Evolution at a Crossroads: The New Biology and the New Philosophy of Science*, edited by D. J. Depew and B. H. Weber, 227–60. Cambridge, Mass.: MIT Press.

Depew, D. J., and B. H. Weber. 1994. *Darwinism Evolving*. Cambridge, Mass.: MIT Press.

Desmond, A. 1982. *Archetypes and Ancestors: Paleontology in Victorian Britain, 1850–1875*. London: Blond and Briggs.

Diderot, D. 1943. *Diderot: Interpreter of Nature*. New York: International Publishers.

Dmitriev, V. Yu., and A. G. Ponomarenko. 2002. General features of insect history. In *History of Insects,* edited by A. P. Rasnitsyn and D. L. J. Quicke, 325–435. Dordrecht: Kluwer.

Dobzhansky, T. 1937. *Genetics and the Origin of Species*. New York: Columbia University Press.

Doolittle, R. F. 1997. A delicate balance. *Boston Review* 22, no. 1: 28–9.

Dreiser, T. 1900. *Sister Carrie*. New York: Doubleday, Page.

Dudley, J. W. 1977. Seventy-six generations of selection for oil and protein percentages in maize. In *Proceedings of the International Conference on Quantitative Genetics,* edited by E. Pollak, O. Kempthorne, and T. B. Bailey, 459–73. Ames: Iowa State University Press.

Edwards, A. W. F. 2003. Human genetic diversity: Lewontin's fallacy. *BioEssays* 25: 798–801.

Engels, F. 1964. *The Dialectics of Nature, 1873–86*, 3rd ed., translated by C. Dutt, edited by D. Amigoni and J. Wallace. Moscow: Progress.

Erskine, F. 1995. "The Origin of Species" and the science of female inferiority. In *Charles Darwin's "The Origin of Species": New Interdisciplinary Essays*, 95–121. Manchester: Manchester University Press.

Falk, D. 2004. *Braindance: New Discoveries about Human Origins and Brain Evolution*. Gainesville: University of Florida Press.

Farley, J. 1977. *The Spontaneous Generation Controversy from Descartes to Oparin*. Baltimore: Johns Hopkins University Press.

Feder, J. L., C. A. Chilcote, and G. L. Bush. 1988. Genetic differentiation between sympatric host races of the apple maggot fly, *Rhagoletis pomonella*. *Nature* 336: 61–4.

Feduccia, A. 1996. *The Origin and Evolution of Birds*. New Haven: Yale University Press.

Fisher, R. A. 1930. *The Genetical Theory of Natural Selection*. Oxford: Oxford University Press.

Fisher, R. A. 1936. Has Mendel's work been rediscovered? *Annals of Science* 1: 115–37.

Fodor, J. 1983. *The Modularity of Mind*. Cambridge, Mass.: MIT Press.

Ford, E. B. 1964. *Ecological Genetics*. London: Methuen.

Fox, S. W. 1988. *The Emergence of Life: Darwinian Evolution from the Inside*. New York: Basic Books.

Frank, R. 1988. *Passions within Reason*. New York: Norton.

Frank, R. 1999. "Cooperation through Emotional Commitment." Unpublished manuscript.

Freeman, S., and J. C. Herron. 2004. *Evolutionary Analysis*, 3rd ed. Englewood Cliffs, N.J.: Prentice-Hall.

Fry, I. 2000. *The Emergence of Life on Earth*. New Brunswick, N.J.: Rutgers University Press.

Gee, H. 1996. Box of bones 'clinches' identity of Piltdown palaeontology hoaxer. *Nature* 382: 261–2.

Gee, H. 2004. Flores, God, and cryptozoology. muse@nature.com, 27 October 2004.

Giere, R. 1988. *Explaining Science: A Cognitive Approach*. Chicago: University of Chicago Press.

Gilbert, S. F., J. M. Opitz, and R. A. Raff. 1996. Resynthesizing evolutionary and developmental biology. *Developmental Biology* 173: 357–72.

Gilbert, W. 1986. The RNA world. *Nature* 319: 618.

Gish, D. 1973. *Evolution: The Fossils Say No!* San Diego, Calif.: Creation-Life.

Goodwin, B. 2001. *How the Leopard Changed Its Spots*, 2nd ed. Princeton, N.J.: Princeton University Press.

Gosse, P. 1857. *Omphalos: An Attempt to Untie the Geological Knot*. London: John Van Voorst.

Gould, S. J. 1971. D'Arcy Thompson and the science of form. *New Literary History* 2: 229–58.

Gould, S. J. 1974. The evolutionary significance of "bizarre" structures: antler size and skull size in the "Irish Elk," *Megalocerus giganteus*. *Evolution* 28: 191–220.

Gould, S. J. 1977. *Ontogeny and Phylogeny*. Cambridge, Mass.: Belknap Press.

Gould, S. J. 1980a. Is a new and general theory of evolution emerging? *Paleobiology* 6: 119–30.

Gould, S. J. 1980b. The Piltdown conspiracy. *Natural History* 89 (August): 8–28.

Gould, S. J. 1981. *The Mismeasure of Man*. New York: Norton.

Gould, S. J. 1982. Darwinism and the expansion of evolutionary theory. *Science* 216: 380–7.

Gould, S. J. 1989. *Wonderful Life: The Burgess Shale and the Nature of History*. New York: Norton.

Gould, S. J. 2002. *The Structure of Evolutionary Theory*. Cambridge, Mass.: Harvard University Press.

Gould, S. J., and R. C. Lewontin. 1979. The spandrels of San Marco and the Panglossian paradigm: a critique of the adaptationist program. *Proceedings of the Royal Society of London, Series B: Biological Sciences* 205: 581–98.

Grant, M. 1916. *The Passing of the Great Race, or The Racial Basis of European History*. New York: Charles Scribner's Sons.

Grant, P. R. 1986. *Ecology and Evolution of Darwin's Finches*. Princeton, N.J.: Princeton University Press.

Grant, P. R. 1991. Natural selection and Darwin's finches. *Scientific American* (October): 82–7.

Grant, P. R., and B. R. Grant. 1995. Predicting microevolutionary responses to directional selection on heritable variation. *Evolution* 49: 241–51.

Grant, R. B., and P. R. Grant. 1989. *Evolutionary Dynamics of a Natural Population: The Large Cactus Finch of the Galapagos*. Chicago: University of Chicago Press.

Gray, A. 1876. *Darwiniana*. New York: Appleton.

Gray, A. 1881. *Structural Botany*, 6th ed. London: Macmillan.

Haeckel E. 1866. *Generelle Morphologie der Organismen*. Berlin: Georg Reimer.

Haeckel, E. 1874. *Anthropogenie oder Entwickelungsgeschichte des Menschen*. Leipzig: Engelmann.

Haldane, J. B. S. 1929. The origin of life. *Rationalist Annual*: 1–10.

Haldane, J. B. S. 1949. *What Is Life?* London: Lindsay Drummond.

Hamilton, W. D. 1964a. The genetical evolution of social behaviour I. *Journal of Theoretical Biology* 7: 1–16.

Hamilton, W. D. 1964b. The genetical evolution of social behaviour II. *Journal of Theoretical Biology* 7: 17–32.

Hamilton, W. D., R. Axelrod, and R. Tanese. 1990. Sexual reproduction as an adaptation to resist parasites. *Proceedings of the National Academy of Science, USA* 87, no. 9: 3566–73.

Hartwig, W. C. 2002. *The Primate Fossil Record*. Cambridge: Cambridge University Press.

Hempel, C. G. 1966. *Philosophy of Natural Science*. Englewood Cliffs, N.J.: Prentice-Hall.

Herbert, S. 2005. The Darwinian Revolution revisited. *Journal of the History of Biology* 38: 51–66.

Highfield, R. 1996. Old canvas trunk holds identity of Piltdown hoaxer. *Daily Telegraph*. (London), May 23.

Hillis, D. M., J. P. Huelsenbeck, and C. W. Cunningham. 1994. Application and accuracy of molecular phylogenies. *Science* 264: 671–7.

Hillis, D. M., C. Moritz, and B. K. Mable, editors. 1996. *Molecular Systematics*. Sunderland, Mass.: Sinauer.

Himmelfarb, G. 1959. *Darwin and the Darwinian Revolution*. London: Chatto and Windus.

Hodge, C. 1874. *What Is Darwinism?* New York: Scribner's.

Hodge, M. J. S. 2005. Against "Revolution" and "Evolution." *Journal of the History of Biology* 28: 101–121.

Hoffman, P. F., A. J. Kaufman, G. P. Halverson, and D. P. Schrag. 1998. A neoproterozoic Snowball Earth. *Science* 281: 1342–6.

Holden, C. 2004. The origin of speech. *Science* 303: 1316–19.

Holt, L. E. 1964. *Samuel Butler*. New York: Twayne.

Honderich, T. 1993. *How Free Are You? The Determinism Problem*. Oxford: Oxford University Press.

Hooper, J. 2002. *Of Moths and Men: The Untold Story of Science and the Peppered Moth*. New York: Norton.

Hopson, J. A. 1975. The evolution of cranial display structures in hadrosaurian dinosaurs. *Paleobiology* 1: 21–43.

Houston, A. I., and J. M. McNamara. 1999. *Models of Adaptive Behaviour*. Cambridge: Cambridge University Press.

Howlett, R. J., and M. E. N. Majerus. 1987. The understanding of industrial melanism in the peppered moth (*Biston betularia*) (Lepidoptera: Geometridae). *Biological Journal of the Linnean Society* 30: 31–44.

Hrdy, S. B. 1981. *The Woman that Never Evolved*. Cambridge, Mass.: Harvard University Press.

Hrdy, S. B. 1999. *Mother Nature: A History of Mothers, Infants, and Natural Selection*. New York: Pantheon Books.

Hull, D., editor. 1973. *Darwin and His Critics*. Cambridge, Mass.: Harvard University Press.

Hull, D. 1988. *Science as a Process*. Chicago: University of Chicago Press.

Hull, D. L., and M. Ruse, editors. 1998. *Readings in the Philosophy of Biology: Oxford Readings in Philosophy*. Oxford: Oxford University Press.

Hume, D. 1978. *A Treatise of Human Nature*. Oxford : Oxford University Press.

Huxley, J. S. 1942. *Evolution: The Modern Synthesis*. London: Allen and Unwin.

Huxley, J. S. 1943. *TVA: Adventure in Planning*. London: Scientific Book Club.

Huxley, T. H. 1863. *Evidence as to Man's Place in Nature*. London: Williams and Norgate.

Huxley, T. H. 1877. *American Addresses, with a Lecture on the Study of Biology*. London: Macmillan.

Huxley, T. H. 1893. *Collected Essays. Darwiniana*. London: Macmillan.

Huxley, T. H. 1988. Evolution and ethics. In *Evolution and Ethics: T. H. Huxley's 'Evolution and Ethics' with New Essays on its Victorian and Sociobiological Context*, edited by J. C. Williams, and J. Paradis, 57–174. Princeton, N.J.: Princeton University Press.

Irwin, D. E., S. Bensch, and T. D. Price. 2001. Speciation in a ring. *Nature* 409: 333–7.

Isaac, G. 1983. Aspects of human evolution. In *Evolution from Molecules to Men*, edited by D. S. Bendall, 509–43. Cambridge: Cambridge University Press.

Jablonski, N. G., and G. Chaplin. 2003. Skin deep. *Scientific American* 13, no. 2: 72–9.

Jacob, F. 1977. Evolution and tinkering. *Science* 196: 1161–6.

Johanson, D., and M. Edey. 1981. *Lucy. The Beginnings of Humankind*. New York: Simon and Schuster.

John Paul II. 1997. The Pope's message on evolution. *Quarterly Review of Biology* 72: 377–83.

Johnson, P. E. 1991. *Darwin on Trial*. Washington, D.C.: Regnery Gateway.

Johnson, P. E. 1995. *Reason in the Balance: The Case against Naturalism in Science, Law and Education*. Downers Grove, Ill.: InterVarsity Press.

Johnson, P. E. 1997. *Defeating Darwinism by Opening Minds*. Downers Grove, Ill.: InterVarsity Press.

Jones, J. S. 2001. *Darwin's Ghost: The* **Origin of Species** *Updated*. New York: Ballantine Books.

Kant, I. 1959. *Foundations of the Metaphysics of Morals*. Indianapolis: Bobbs-Merrill.

Kauffman, S. A. 1993. *The Origins of Order: Self-Organization and Selection in Evolution*. Oxford: Oxford University Press.

Kauffman, S. A. 1995. *At Home in the Universe: The Search for the Laws of Self-Organization and Complexity*. New York: Oxford University Press.

Kettlewell, H. B. D. 1973. *The Evolution of Melanism*. Oxford: Clarendon.

Kimura, M. 1983. *Neutral Theory of Molecular Evolution*. Cambridge: Cambridge University Press.

Kitcher, P. 2001. Born-again Creationism. In *Intelligent Design Creationism and Its Critics: Philosophical, Theological, and Scientific Perspectives*, edited by R. T. Pennock, 257–88. Cambridge, Mass.: MIT Press.

Kitcher, P. 2003. *In Mendel's Mirror: Philosophical Reflections on Biology*. New York: Oxford University Press.

Knoll, A. 2003. *Life on a Young Planet: The First Three Billion Years of Evolution on Earth*. Princeton, N.J.: Princeton University Press.

Knoll, A., and S. B. Carroll. 1999. Early animal evolution: emerging views from comparative biology and geology. *Science* 284: 2129–37.
Kropotkin, P. [1902] 1955. *Mutual Aid.* Boston: Extending Horizons Books.
Kuhn, T. 1962. *The Structure of Scientific Revolutions*. Chicago: University of Chicago Press.
Lamarck, J. B. 1809. *Philosophie zoologique*. Paris: Dentu.
Leakey, M., and A. Walker. 2003. Early hominid fossils from Africa. *Scientific American* 13, no. 2: 14–19.
Lennox, J. G. 2005. Darwin's methodological evolution. *Journal of the History of Biology* 38: 85–99.
Leonard, W. R. 2003. Food for thought: dietary change was a driving factor in human evolution. *Scientific American* (January), 62–71.
Leonard, W. R., and M. L. Robertson. 1997. Rethinking the energetics of bipedality. *Current Anthropology* 38, no. 2: 304–9.
Lewontin, R. C. 1978. Adaptation. *Scientific American* 239, no. 3: 213–30.
Lewontin, R. C. 1982. *Human Diversity*. New York: Scientific American Library.
Lewontin, R. C. 1991. *Biology as Ideology: The Doctrine of DNA.* Toronto: Anansi.
Lewontin, R. C., J. A. Moore, W. B. Provine, and B. Wallace, editors. 1981. *Dobzhansky's Genetics of Natural Populations I–XLIII.* New York: Columbia University Press.
Lewontin, R. C., Steven Rose, and Leon J. Kamin. 1984. *Not in Our Genes: Biology, Ideology and Human Nature*. New York: Pantheon.
Lieberman, P. 1984. *The Biology and Evolution of Language*. Cambridge, Mass.: Harvard University Press.
Lipton, P. 1991. *Inference to the Best Explanation*. London: Routledge.
Lumsden, C. J., and E. O. Wilson. 1981. *Genes, Mind, and Culture*. Cambridge, Mass.: Harvard University Press.
Lyell, C. 1830–33. *Principles of Geology: Being an Attempt to Explain the Former Changes in the Earth's Surface by Reference to Causes Now in Operation*. London: John Murray.
Mackie, J. 1977. *Ethics*. Harmondsworth, Mddx.: Penguin.
Mackie, J. 1979. *Hume's Moral Theory*. London: Routledge and Kegan Paul.
Macnair, M. R., and P. Christie. 1983. Reproductive isolation as a pleiotropic effect of copper tolerance in *Mimulus guttatus*. *Heredity* 50: 295–302.
Majerus, M. E. N. 1998. *Melanism: Evolution in Action*. Oxford: Oxford University Press.
Mangelsdorf, P. C. 1974. *Corn: Its Origin, Evolution, and Improvement*. Cambridge, Mass.: Harvard University Press.
Margulis, L. 1970. *Origin of Eukaryotic Cells*. New Haven, Conn.: Yale University Press.
Margulis, L. 1993. *Symbiosis in Cell Evolution*. New York: Freeman.
Maynard Smith, J. 1969. The status of neo-Darwinism. In *Towards a Theoretical Biology*, edited by C. H. Waddington. Edinburgh: Edinburgh University Press.
Maynard Smith, J. 1978a. The evolution of behavior. *Scientific American* 239, no. 3: 176–93.

Maynard Smith, J. 1978b. *The Evolution of Sex*. Cambridge: Cambridge University Press.

Maynard Smith, J. 1982. *Evolution and the Theory of Games*. Cambridge: Cambridge University Press.

Maynard Smith, J., R. Burian, S. Kauffman, P. Alberch, J. Campbell, B. Goodwin, R. Lande, D. Raup, and L. Wolpert. 1985. Developmental constraints and evolution. *Quarterly Review of Biology* 60: 265–87.

Mayr, E. 1942. *Systematics and the Origin of Species*. New York: Columbia University Press.

Mayr, E. 1954. Change of genetic environment and evolution. In *Evolution as a Process*, edited by J. S. Huxley, A. C. Hardy, and E. B. Ford, 157–80. London: Allen and Unwin.

Mayr, E. 1963. *Animal Species and Evolution*. Cambridge, Mass.: Harvard University Press.

McDonald, J. H., G. K. Chambers, J. David, and F. J. Ayala. 1977. Adaptive response due to changes in gene regulation: a study with *Drosophila*. *Proceedings of the National Academy of Sciences USA* 74: 4562–6.

McDonald, J. H., and M. Kreitman. 1991. Adaptive protein evolution at the Adh locus in *Drosophila*. *Nature* 351: 652–4.

McEwan, I. 1997. *Enduring Love*. London: Cape.

McMullin, E. 1983. Values in science. In *PSA 1982*, edited by P. D. Asquith and T. Nickles, 3–28. East Lansing, Mich.: Philosophy of Science Association.

McMullin, E., editor. 1985. *Evolution and Creation*. Notre Dame, Ind.: University of Notre Dame Press.

McMullin, E. 1991. Plantinga's defense of special creation. *Christian Scholar's Review* 21, no. 1: 55–79.

McShea, D. W. 1991. Complexity and evolution: what everybody knows. *Biology and Philosophy* 6, no. 3: 303–25.

Meléndez-Hevia, E., T. G. Waddell, and M. Cascante. 1996. The puzzle of the Krebs citric acid cycle: assembling the pieces of chemically feasible reactions, and opportunism in the design of metabolic pathways during evolution. *Journal of Molecular Evolution* 43: 293–303.

Miller, K. 1999. *Finding Darwin's God*. New York: Harper and Row.

Miller, S. L. 1953. A production of amino acids under possible primitive Earth conditions. *Science* 117: 528–9.

Mitchison, G. J. 1977. Phyllotaxis and the Fibonacci series. *Science* 196: 270–5.

Mithen, S. 1996. *The Prehistory of the Mind*. London: Thames and Hudson.

Moore, A. 1890. The Christian doctrine of God. In *Lux Mundi*, edited by C. Gore. London: John Murray: 41–81.

Morewood, M. J., and others. 2004. Archaeology and age of a new hominin from Flores in eastern Indonesia. *Nature* 431: 1087–91.

Mullin, R. B. 1996. *Miracles and the Modern Religious Imagination*. New Haven, Conn.: Yale University Press.

Murphy, J. 1982. *Evolution, Morality, and the Meaning of Life*. Totowa, N.J.: Rowman and Littlefield.

Nagel, E. 1961. *The Structure of Science: Problems in the Logic of Scientific Explanation*. New York: Harcourt, Brace and World.

Newman, J. H. 1973. *The Letters and Diaries of John Henry Newman, XXV*. edited by C. S. Dessain, and T. Gornall. Oxford: Clarendon Press.

Niklas, K. J. 1988. The role of phyllotactic pattern as a 'developmental constraint' on the interception of light by leaf surfaces. *Evolution* 42: 1–16.

Niklas, K. J. 1997. *The Evolutionary Biology of Plants*. Chicago: University of Chicago Press.

Norris, F. 1901. *Octopus: A Story of California*. New York: Doubleday, Page.

Nozick, R. 1981. *Philosophical Explanations*. Cambridge, Mass.: Harvard University Press.

Nyhart, L. K. 1995. *Biology Takes Form: Animal Morphology and the German Universities*. Chicago: University of Chicago Press.

Oparin, A. [1924] 1967. The origin of life. In his *The Origin of Life*, translated by A. Synge, 199–234. Cleveland: World.

Orgel, L. E. 1994. The origin of life on the Earth. *Scientific American* 271, no. 10: 77–83.

Orgel, L. E. 1998. The origin of life – a review of facts and speculations. *Trends in Biochemical Sciences* 23: 491–500.

Orzack, S. H., and E. Sober. 1994. Optimality models and the test of adaptationism. *American Naturalist* 143: 361–80.

Orzack, S. H., and E. Sober., editors. 2001. *Adaptationism and Optimality*. Cambridge: Cambridge University Press.

Ospovat, D. 1981. *The Development of Darwin's Theory: Natural History, Natural Theology, and Natural Selection, 1838–1859*. Cambridge: Cambridge University Press.

Oster, G., and E. O. Wilson. 1978. *Caste and Ecology in the Social Insects*. Princeton, N.J.: Princeton University Press.

Pace, N., and T. Marsh. 1986. RNA catalysis and the origin of life. *Origins of Life* 16: 97–116.

Paley, W. [1802] 1819. *Natural Theology* (*Collected Works*, vol. 4). London: Rivington.

Pannenberg, W. 1993. *Towards a Theology of Nature*. Louisville, Ky.: Westminster/John Knox Press.

Paterniani, E. 1969. Selection for reproductive isolation between two populations of maize, *Zea mays L. Evolution* 23: 534–47.

Pennisi, E. 2004. The first language? *Science* 303: 1319–20.

Pennock, R. 1998. *Tower of Babel: Scientific Evidence and the New Creationism*. Cambridge, Mass.: MIT Press.

Pickford, M. B. Senut, D. Gommercy, and J. Treil. 2002. Bipedalism in *Orrorin tugenensis* revealed by its femora. Compte Rendus: *Palevol* 1, no. 4: 191–203.

Pinker, S. 1994. *The Language Instinct: How the Mind Creates Language*. New York: William Morrow.

Pinker, S. 1997. *How the Mind Works*. New York: Norton.

Pinker, S. 2002. *The Blank Slate: The Modern Denial of Human Nature*. London: Allen Lane.

Plantinga, A. 1983. Reason and belief in God. In *Faith and Rationality: Reason and Belief in God*, edited by A. Plantinga and N. Wolterstorff, 16–93. Notre Dame, Ind.: University of Notre Dame Press.

Plantinga, A. 1991. An evolutionary argument against naturalism. *Logos* 12: 27–49.

Plantinga, A. 1991. When faith and reason clash: evolution and the Bible. *Christian Scholar's Review* 21, no. 1: 8–32. [Reprinted in D. Hull and M. Ruse, eds., *The Philosophy of Biology* (Oxford: Oxford University Press, 1998), 674–97.]

Plantinga, A. 1994. "Naturalism Defeated." Unpublished manuscript.

Plantinga, A. 1997. Methodological naturalism. *Perspectives on Science and Christian Faith* 49, no. 3: 143–54.

Popper, K., and J. Eccles. 1977. *The Self and Its Brain*. Berlin: Springer.

Popper, K. R. 1959. *The Logic of Scientific Discovery*. London: Hutchinson.

Popper, K. R. 1974. Darwinism as a metaphysical research programme. In *The Philosophy of Karl Popper*, edited by P. A. Schilpp, vol. 1: 133–43. LaSalle, Ill.: Open Court.

Provine, W. B. 1971. *The Origins of Theoretical Population Genetics*. Chicago: University of Chicago Press.

Quine, W. V. O. 1969. *Ontological Relativity and Other Essays*. New York: Columbia University Press.

Raff, R. 1996. *The Shape of Life: Genes, Development, and the Evolution of Animal Form*. Chicago: University of Chicago Press.

Rawls, J. 1971. *A Theory of Justice*. Cambridge, Mass.: Harvard University Press.

Reeve, H. K., and L. Keller. 1999. Levels of selection: burying the units-of-selection debate and unearthing the crucial new issues. In *Levels of Selection in Evolution*, edited by L. Keller, 3–14. Princeton, N.J.: Princeton University Press.

Reeve, H. K., and P. W. Sherman. 1993. Adaptation and the goals of evolutionary research. *Quarterly Review of Biology* 68: 1–32.

Reichenbach, B. R. 1976. Natural evils and natural laws: a theodicy for natural evil. *International Philosophical Quarterly* 16: 179–96.

Reichenbach, B. R. 1982. *Evil and a Good God*. New York: Fordham University Press.

Rice, W. R., and G. W. Salt. 1990. The evolution of reproductive isolation as a correlated character under sympatric conditions: experimental evidence. *Evolution* 44: 1140–52.

Richards, R. J. 1987. *Darwin and the Emergence of Evolutionary Theories of Mind and Behavior*. Chicago: University of Chicago Press.

Richards, R. J. 1992. *The Meaning of Evolution: The Morphological Construction and Ideological Reconstruction of Darwin's Theory*. Chicago: University of Chicago Press.

Richards, R. J. 2003. *The Romantic Conception of Life: Science and Philosophy in the Age of Goethe*. Chicago: University of Chicago Press.

Richardson, M. K., J. Hanken, L. Selwood, G. M. Wright, R. J. Richards, and C. Pieae. 1998. Haeckel, embryos, and evolution. *Science* 280: 983–6.

Ridley, M. 1986. *Evolution and Classification: The Reformation of Cladism*. New York: Longman.

Rogers, A. R., and A. Mukherjee. 1992. Quantitative genetics of sexual dimorphism in human body size. *Evolution* 46: 226–34.

Rolston III, H. 1987. *Science and Religion*. New York: Random House.

Rosenberg, N. A., J. K. Pritchard, J. L. Weber, H. M. Cann, K. K. Kidd, L. A. Zhivotovsky, and M. Feldman. 2002. Genetic structure of human populations. *Science* 298: 2381–5.

Ross, R. H., editor. 1973. *Alfred, Lord Tennyson: "In Memoriam": An Authoritative Text, Backgrounds and Sources of Criticism*. New York: Norton.

Roth, G., J. Blanke, and D. B. Wake. 1994. Cell size predicts morphological complexity in the brains of frogs and salamanders. *Proceedings of the National Academy of the Sciences, USA* 91: 4796–800.

Rudwick, M. J. S. 1972. *The Meaning of Fossils*. New York: Science History Publications.

Ruse, M. 1973. *The Philosophy of Biology*. London: Hutchinson.

Ruse, M. 1982. *Darwinism Defended: A Guide to the Evolution Controversies*. Reading, Mass.: Benjamin/Cummings.

Ruse, M. 1987. Darwin and determinism. *Zygon* 22: 419–42.

Ruse, M., editor. 1988a. *But Is It Science? The Philosophical Question in the Creation/Evolution Controversy*. Buffalo, N.Y.: Prometheus.

Ruse, M. 1988b. *Homosexuality: A Philosophical Inquiry*. Oxford: Blackwell.

Ruse, M. 1989. *The Darwinian Paradigm: Essays on Its History, Philosophy and Religious Implications*. London: Routledge.

Ruse, M. 1996a. The Darwin Industry: a guide. *Victorian Studies* 39, no. 2: 217–35.

Ruse, M. 1996b. *Monad to Man: The Concept of Progress in Evolutionary Biology*. Cambridge, Mass.: Harvard University Press.

Ruse, M. 1998. *Taking Darwin Seriously: A Naturalistic Approach to Philosophy*, 2nd ed. Buffalo, N.Y.: Prometheus.

Ruse, M. 1999a. *The Darwinian Revolution: Science Red in Tooth and Claw*, 2nd ed. Chicago: University of Chicago Press.

Ruse, M. 1999b. *Mystery of Mysteries: Is Evolution a Social Construction?* Cambridge, Mass.: Harvard University Press.

Ruse, M. 2000. *The Evolution Wars: A Guide to the Controversies*. Santa Barbara, Calif.: ABC-CLIO.

Ruse, M. 2001. *Can a Darwinian Be a Christian? The Relationship between Science and Religion*. Cambridge: Cambridge University Press.

Ruse, M. 2003. *Darwin and Design: Does Evolution Have a Purpose?* Cambridge, Mass.: Harvard University Press.

Ruse, M. 2005a. The Darwinian Revolution as seen in 1979 and as seen twenty-five years later in 2004. *Journal of the History of Biology* 38: 3–17.

Ruse, M. 2005b. *The Evolution-Creation Struggle*. Cambridge, Mass.: Harvard University Press.

Ruse, M., and D. Castle, editors. 2002. *Genetically Modified Foods*. Buffalo, N.Y.: Prometheus.

Russell, E. S. 1916. *Form and Function: A Contribution to the History of Animal Morphology*. London: John Murray.

Russett, C. E. 1976. *Darwin in America: The Intellectual Response, 1865–1912*. San Francisco: Freeman.

Ruvolo, M. 1994. Molecular evolutionary processes and conflicting gene trees: the Hominoid case. *American Journal of Physical Anthropology* 94: 89–113.

Ruvolo, M. 1995. Seeing the forest and the trees: replies to Marks, Rogers and Commuzzie, Green and Djian. *American Journal of Physical Anthropology* 98: 218–32.

Sarich, V., and A. C. Wilson. 1967. Immunological time scale for hominid evolution. *Science* 158: 1200–3.

Secord, J. A. 2000. *Victorian Sensation: The Extraordinary Publication, Reception, and Secret Authorship of Vestiges of the Natural History of Creation*. Chicago: University of Chicago Press.

Segerstrale, U. 1986. Colleagues in conflict: an in vitro analysis of the sociobiology debate. *Biology and Philosophy* 1: 53–88.

Sepkoski Jr., J. J. 1978. A kinetic model of Phanerozoic taxonomic diversity I: analysis of marine orders. *Paleobiology* 4: 223–51.

Sepkoski Jr., J. J. 1979. A kinetic model of Phanerozoic taxonomic diversity II: early Paleozoic families and multiple equilibria. *Paleobiology* 5: 222–52.

Sepkoski Jr., J. J. 1984. A kinetic model of Phanerozoic taxonomic diversity III: post-Paleozoic families and mass extinctions. *Paleobiology* 10: 246–67.

Serpell, J., editor. 1995. *The Domestic Dog: Its Evolution, Behaviour, and Interactions with People*. Cambridge: Cambridge University Press.

Shaw, G. B. 1988. *Back to Methuselah. A Metabiological Pentateuch*. Harmondsworth, Mddx.: Penguin.

Shermer, M. 2002. *In Darwin's Shadow: The Life and Science of Alfred Russel Wallace*. New York: Oxford University Press.

Shipman, P. 2002. *The Man who Found the Missing Link: Eugene Dubois and His Lifelong Quest to Prove Darwin Right*. Cambridge, Mass.: Harvard University Press.

Shor, E. N. 1974. *The Fossil Feud between E. D. Cope and O. C. Marsh*. Hicksville, N.Y.: Exposition Press.

Simpson, G. G. 1944. *Tempo and Mode in Evolution*. New York, N.Y.: Columbia University Press.

Simpson, G. G. 1949. *The Meaning of Evolution*. New Haven, Conn.: Yale University Press.

Simpson, G. G. 1951. *Horses*. New York: Oxford University Press.

Skyrms, B. 1998. *Evolution of the Social Contract*. Cambridge: Cambridge University Press.

Skyrms, B. 2002. Game theory, rationality and evolution of the social contract. In *Evolutionary Origins of Morality*, edited by L. D. Katz, 269–84. Bowling Green, Ohio: Imprint Academic.

Smocovitis, V. B. 1999. The 1959 Darwin centennial celebration in America. *Osiris* 14: 274–323.

Sober, E. 1984. *The Nature of Selection*. Cambridge, Mass.: MIT Press.

Sober, E. 1988. *Reconstructing the Past: Parsimony, Evolution, and Inference*. Cambridge, Mass.: MIT Press.

Sober, E., editor. 1994. *Conceptual Issues in Evolutionary Biology*, 2nd ed. Cambridge, Mass.: MIT Press.

Sober, E., and D. S. Wilson. 1997. *Unto Others: The Evolution of Altruism*. Cambridge, Mass.: Harvard University Press.

Spencer, F. 1990. *Piltdown: A Scientific Forgery*. New York: Oxford University Press.

Spencer, H. 1851. *Social Statics; Or the Conditions Essential to Human Happiness Specified and the First of them Developed*. London: J. Chapman.

Spencer, H. 1857. Progress: its law and cause. *Westminster Review* 67: 244–67.

Stapledon, W. O. 1930. *Last and First Men: A Story of the Near and Far Future*. London: Methuen.

Stebbins, G. L. 1950. *Variation and Evolution in Plants*. New York: Columbia University Press.

Stebbins, G. L., and F. J. Ayala. 1981. Is a new evolutionary synthesis necessary? *Science* 213: 967–71.

Sterelny, K. 2003. *Thought in a Hostile World: The Evolution of Human Cognition*. Oxford: Blackwell.

Sterelny, K., and P. E. Griffiths. 1999. *Sex and Death: An Introduction to Philosophy of Biology*. Chicago: University of Chicago Press.

Stinson, S., B. Bogin, R. Huss-Ashmore, and D. O'Rourke, editors. 2000. *Human Biology: An Evolutionary and Biocultural Approach*. New York: Wiley-Liss.

Swinburne, R. G. 1970. *The Concept of Miracle*. London: Macmillan.

Syvanen, M., and C. I. Kado, editors. 2002. *Horizontal Gene Transfer*, 2nd ed. London: Academic Press.

Szostak, J. W., D. P. Bartel, and P. L. Luisi. 2001. Synthesizing life. *Nature* 409: 387–90.

Tattersall, I. 2003a. Out of Africa again . . . and again. *Scientific American* 13, no. 2: 38–45.

Tattersall, I. 2003b. Once we were not alone. *Scientific American* 13, no. 2: 21–27.

Thewissen, J. G. M., S. T. Hussain, and M. Arif. 1994. Fossil evidence for the origin of aquatic locomotion in archaeocete whales. *Science* 263: 210–12.

Thompson, D. W. 1948. *On Growth and Form*, 2nd ed. Cambridge: Cambridge University Press.

Thompson, R. P. 1989. *The Structure of Biological Theories*. Albany, N.Y.: State University of New York Press.

Thornhill, R., and C. T. Palmer. 2001. *A Natural History of Rape: Biological Bases of Sexual Coercion*. Cambridge, Mass.: Bradford Books.

Tishkoff, S. A., E. Dietzsch, W. Speed, A. J. Pakstis, J. R. Kidd, K. Cheung, B. Bonne-Tamir, A. S. Santachiara-Benerecetti, P. Moral, M. Krings, S. Paabo, E. Watson, N. Risch., T. Jenkins, and K. K. Kidd. 1996. Global patterns of linkage disequilibrium at the CD4 locus and modern human origins. *Science* 271: 1380–7.

Toulmin, S. 1967. The evolutionary development of science. *American Scientist* 57: 456–71.

Travis, C. B., editor. 2003. *Evolution, Gender, and Rape*. Cambridge, Mass.: MIT Press.

Trivers, R. L. 1971. The evolution of reciprocal altruism. *Quarterly Review of Biology* 46: 35–57.

Tutt, J. W. 1891. *Melanism and Melanochroism in British Lepidoptera*. London: Swan Sonnenschein.

Vogel, S. 1988. *Life's Devices: The Physical World of Animals and Plants*. Princeton, N.J.: Princeton University Press.

von Berhardi, F. 1912. *Germany and the Next War*. London: Edward Arnold.

Wachtershauser, G. 1992. Groundwork for an evolutionary biochemistry: the non-sulfur world. *Progress in Biophysics and Molecular Biology* 58: 85–201.

Wagar, W. 1972. *Good Tidings: The Belief in Progress from Darwin to Marcuse*. Bloomington: Indiana University Press.

Wallace, A. R. 1900. *Studies: Scientific and Social*. London: Macmillan.

Ward, K. 1996. *God, Chance and Necessity*. Oxford: Oneworld.

Watson, J. 1968. *The Double Helix*. New York: Signet Books.

Wavell, S., and W. Iredale. 2004. Sorry, says atheist-in-chief, I do believe in God after all. *The Sunday Times* (London), December 12, sec. 1, p. 7.

Weiner, J. S. 1955. *The Piltdown Forgery*. New York: Oxford University Press.

Weishampel, D. B. 1981. Acoustic analyses of potential vocalization in lambeosaurine dinosaurs (Reptilia: Ornithischia). *Paleobiology* 7: 252–61.

Weishampel, D. B. 1997. Dinosaurian cacophony. *BioScience* 47, no. 3: 150–8.

Weldon, W. F. R. 1898. Presidential address to the Zoological Section of the British Association. *Transactions of the British Association*, 887–902.

Wells, H. G. 1895. *The Time Machine*. London: Heinemann.

Wells, J. 2000. *Icons of Evolution: Science or Myth?* Washington, D.C.: Regnery.

Werner, E. E., and D. J. Hall. 1988. Ontogenetic habitat shifts in bluegill: the foraging rate-predation risk trade-off. *Ecology* 69: 1352–66.

West, S. A. 2002. The evolution of sex. In *Encyclopedia of Evolution*, edited by M. Pagel, vol. 2: 1022–30. Oxford: Oxford University Press.

Whewell, W. 1840. *The Philosophy of the Inductive Sciences* London: Parker.

Whitehead, A. N. 1967. *Science and the Modern World*. New York: The Free Press.

Wilson, E. O. 1975. *Sociobiology: The New Synthesis*. Cambridge, Mass.: Harvard University Press.

Wilson, E. O. 1978. *On Human Nature*. Cambridge, Mass.: Cambridge University Press.

Wilson, E. O. 1980a. Caste and division of labor in leaf cutter ants (hymenoptera formicidae, Atta) I: the overall pattern in Atta sexdens. *Behavioral Ecology and Sociobiology* 7: 143–56.

Wilson, E. O. 1980b. Caste and division of labor in leaf cutter ants (hymenoptera formicidae, Atta) II: the ergonomic optimization of leaf cutting. *Behavioral Ecology and Sociobiology* 7: 157–65.

Wilson, E. O. 1983a. Caste and division of labor in leaf cutter ants (hymenoptera formicidae, Atta) III: ergonomic resiliency in foraging by Atta cephalotes. *Behavioral Ecology and Sociobiology* 14: 47–54.

Wilson, E. O. 1983b. Caste and division of labor in leaf cutter ants (hymenoptera formicidae, Atta) IV: colony ontogeny of Atta cephalotes. *Behavioral Ecology and Sociobiology* 14: 55–60.

Wilson, E. O. 1984. *Biophilia*. Cambridge, Mass.: Harvard University Press.

Wilson, E. O. 1992. *The Diversity of Life*. Cambridge, Mass.: Harvard University Press.

Wilson, E. O. 1994. *Naturalist*. Washington, D.C.: Island Books/Shearwater Books.

Wilson, E. O. 2002. *The Future of Life*. New York: Vintage.

Wittgenstein, L. 1923. *Tractatus Logico-Philosophicus*. London: Routledge and Kegan Paul.

Wolpert, L. 1991. *The Triumph of the Embryo*. Oxford: Oxford University Press.

Wolpoff, M., and R. Caspari. 1997. *Race and Human Evolution*. Boulder, CO: Westview.

Wong, K. 2003a. An ancestor to call our own. *Scientific American* 13, no. 2: 4–13.

Wong, K. 2003b. Who were the Neanderthals? *Scientific American* 13, no. 2: 28–37.

Wray, G. A. 2001. Dating branches on the tree of life using DNA. *Genome Biology* 3, no. 1: 0001.1–0001.7.

Wright, C. 1871. *Darwinism: Being an Examination of Mr. St. George Mivart's 'Genesis of Species'*. London: J. Murray.

Wright, S. [1931] 1986. Evolution in Mendelian populations. In *Evolution: Selected Papers*, edited by W. B. Provine, 98–160. Chicago: University of Chicago Press.

Wright, S. [1932] 1986. The roles of mutation, inbreeding, crossbreeding and selection in evolution. In *Evolution: Selected Papers*, edited by W. B. Provine, 161–71. Chicago: University of Chicago Press.

Index

Page references to illustrations appear in italics.